# Troubleshooting Tactics

## How Process Operators Make Critical Decisions

Ron Besuijen

FIRST EDITION

Graphics created by author unless otherwise noted.
Brain in cover is from istockphoto - Naeblys.[1]

Printed by Kindle Direct Publishing

# Table of Contents

# Foreword

This is an unusual book by an unusual author, and it is a privilege for me to provide some background on both the book and the author. But before I do, I want to prepare readers for what you are about to encounter — a thoroughly enjoyable book on what can be a very dry topic, and an experience that can have great potential value to you.

Besuijen has written this book for panel operators. It is hard enough to keep a highly complicated petrochemical unit working efficiently, but the tension really mounts when things start to go wrong and you're not sure why. You have to figure out the problem in order to correct it or at least manage it safely, and you don't have much time. This is the realm of troubleshooting. "Trouble Shooting Tactics: How Process Operators Make Critical Decisions" should be of immense value to you in handling these kinds of challenges and should give you some important reassurance that you can get the job done. The book won't make you an expert like Besuijen — no book can do that — but it can boost your ability and set you on the road to expanding your expertise.

When I wrote that Besuijen was an unusual author, I was not carelessly tossing out a compliment simply because that's what is expected in a Foreword. I meant that sincerely. First of all, Besuijen is not the kind of person you would expect to write a book. He is not an academic. He is not out to pad his resume. He wrote this book because he has immense experience and sincerely wants to help others become more successful. Second, he is a grunt. He knows this stuff from the ground up, if you will allow me to use a cliché except that in this case it isn't a cliché. Besuijen worked on the ground to actually construct some of the units he later operated. Then he was an outside Operator, making sure the units ran well. And then he was a Panel Operator, performing the kind of work that readers currently do. He stressed about keeping the output moving. He stressed about each start-up episode, worrying about what might go wrong this time. He stressed about unpleasant surprises and malfunctions that needed to be diagnosed. He writes from his own personal and professional experience.

Third, he is what we call a "Subject Matter Expert." A person that others respect and call upon for advice when they get stuck.

Fourth, he is a trainer. He has spent many years training panel operators to come up to speed, training them to become better Troubleshooters. He helped design the simulation facility in his plant so that he could provide better training. So, he has already been working hard to prepare Panel Operators to become better decision makers.

Fifth, on top of all that, Besuijen is a researcher. He is a natural researcher, trying out different things and gauging what works. He is also an acknowledged researcher. Through the years he has worked with the Center for Operator Performance to collaborate with credentialed researchers carrying out complex experiments. Besuijen didn't just advise them. He collaborated with them on how to design their studies, how to collect their data, and how to analyze their findings. He did this for many different projects over the course of several years, he deserves an honorary degree in research design, in my opinion.

Sixth, and most important, Besuijen has a very different mindset than you will encounter from others in his profession. He certainly has an engineering mindset. But he also has a cognitive mindset. He is thinking about the people operating the system as well as the system itself. He knows how these systems are constructed and he also knows how the Operators are gaining insights or are getting confused. He can see inside the heads of the decision makers as clearly as he can see inside the workings of the various towers and heating elements and valve overrides. That's what made him such a gifted trainer, and what has made him such a gifted writer.

When I wrote that this is an unusual book by an unusual author, I was thinking about each of these six factors. I was also thinking about the blend of the factors. You, the reader, are unlikely to have met anyone who possesses most of their qualities. And it is the blend of these qualities that is perhaps the most unusual characteristic of the author and the book.

I should also say a little about my own history with Besuijen. *(And by the way, his last name is Dutch and is pronounced "bez-EYE-yon" — it took me a bit to work this out. He kept saying he didn't mind as I kept mispronouncing it, but it bothered me.)* We met about ten years ago through the Center for Operator Performance research program I mentioned above. The COP was holding a semi-annual meeting in Dayton, Ohio. My company was going to get funding for a study to identify the cognitive requirements of panel operators. Besuijen was there as a practitioner, and he volunteered to host our study. My colleague Joseph Borders and I were delighted to accept his offer. We were less delighted when we showed up to do our cognitive interviews at his facility. It was located in Red Deer, Alberta, Canada and we arrived in mid-February. Upon arrival, Joseph and I were asked if we would like to do a brief tour of the plant and that sounded like a good idea except our tour was outside, and it was -11 degrees F., with 15mph winds. Oh, and a thin plastic safety hat in place of the fur-lined cap I had brought with me, and thin thermal gloves in place of the heavily insulated gloves that didn't meet safety standards.

Would I trust Besuijen with my own safety? I did that day as I climbed several stories on a metal ladder, buffeted by the wind, hands and feet getting numb.

In the following years, I have worked with Besuijen on a number of projects: turning our cognitive interviews into decision training exercises, preparing a cognitive debriefing post-incident review guide, developing a program to train-the-trainers to do decision skills development, preparing a workshop to strengthen diagnosis skills, and expanding existing accounts of mental models. In my opinion, these have all been very successful projects. But I still remember the first one. Not just because of the frozen tour that I took upon arrival. What I remember most vividly is how Besuijen quickly mastered the decision scenario concept we were using and without any coaching developed his own scenario — one that was better than the scenarios Borders and I had developed. Borders and I had never seen anyone master the scenario design technique so quickly. That's when, in my mind, Besuijen transformed from an advisor to a colleague.

Now you can see why I am so enthusiastic about this book. It is written to help you become a better decision maker and Troubleshooter, but from the inside, from someone who has been in your shoes and at the same time, someone who brings a research perspective and a cognitive perspective that you won't find elsewhere. The book covers several critical topics: decision making, sensemaking, insight, diagnosis and also physiology and stress and emotional reactions that will inevitably accompany your troubleshooting efforts. I should also mention that the book contains many exciting and informative case studies, drawn from Besuijen's own experiences as well as the experiences of others.

Putting all these together, Trouble Shooting Tactics is a unique, valuable, and thought-provoking contribution that will change how you view troubleshooting in the future, and how you respond to situations that call for calm and effective troubleshooting performance.

**Dr. Gary A Klein**

# Introduction

Process Operators may have to make decisions that can have life altering consequences. These could involve millions of dollars in equipment damage or may involve millions of dollars in production losses. When things go wrong Operators are expected to minimize the damage and to do this in a timely fashion. As an Operator, it has been my experience that we do not do enough to prepare our Operators for these types of decisions.

I have been in the petrochemical business for 33 years--15 of those years on shift operating, including the commissioning and start-up of a major petrochemical facility. For the last 18 years I have been training Operators on high-fidelity simulators.

Sometimes in life our failures are our best teachers and motivators. Shortly after the start-up of the petrochemical facility, we had a control system failure that closed many of the control valves and stopped the flow through the plant resulting in a large amount of flaring. This was compounded by the catalyst in a converter overheating due to the loss of flow.

As the shift Technical Leader, I walked into the control room to pages of alarms and the flare monitor lit up. My training had been to collect all the data and pick the best choice out of several options. If I had been given an hour, I may have come close. We could not have stopped the event as Operators, but we could have limited the impact by shutting down the main compressor sooner, which the Panel Operator did after a few minutes.

I knew I had to do better, but how? I had learned everything I could about the process through all the technical data and emergency procedures. But that was not enough. That event left me feeling defeated.

Fast-forward to reading “Sources of Power”. I read this book with a lot of reflection, and it made me review past decisions I had made--good ones and bad. I concluded that the author Dr. Gary Klein was right on how we make decisions.

Dr. Klein developed the concept of a Recognition Primed Decision to help us understand how we can make critical decisions when there is limited time. This approach helps to free up your thought process and reach the most likely conclusion much faster. These concepts will be discussed in this book in more detail.

I am not writing this book because I am a natural at troubleshooting or at making critical decisions within a limited time. I have some experience problem-solving from my personal experiences, but sometimes I struggled to make timely decisions. I have tried to include everything in this book that I wished I would have known as a new Panel Operator.

Troubleshooting can be broken down into two different approaches. The first approach is to use our reasoning where we gather and analyze all the relevant information. This generates several options and then we develop criteria to prove which one is true. This is the approach that is most often taught and well ingrained from high school science classes.

The second approach is the one we see used when a seasoned Operator approaches a Panel Operator struggling with a problem and can almost immediately tell them what is going on. We may call this using experience or pattern recognition. This is the approach that Operators may rely on when there is limited time available to make a decision.

These are both models that describe how we make decisions. The third approach is trying to merge these two as I believe we likely use both to some extent depending on the situation. When there are time constraints, we may lean more to using the second approach. When there is more time, it can be preferred to evaluate a few options-- especially when the recommendation might involve spending a few million dollars to fix something or having to shut down the process to make a repair. You will want to be as accurate as possible to avoid extended downtime or shutting down the process unnecessarily when there is a simple fix.

I will cover some training techniques that can help develop the skills required for making time-limited decisions (although they will benefit both styles of decision-making). The first is on how to encourage pattern development, and then I will cover scenario-based training exercises.

I have tried to look at troubleshooting from as many different perspectives as possible including managing stress, working with teams, and how to train others to be successful.

This is not intended to be a technical book on how to analyze specific problems for each process. I have included some basic concepts as examples and a general description of control systems.

As you read through this material, I do encourage you to reflect on your past troubleshooting efforts and how you have made decisions in tough situations that you have been through.

I hope you will bear with me while I try to merge these two approaches. I believe this book is the first attempt in our industry, and I hope it is a catalyst to develop other methods to prepare and to make critical decisions in our industry.

**Ron Besuijen**

# CHAPTER 1

# Decision-Making Approaches

We traditionally think that our job as Operators is to perform tasks and follow procedures. During normal routine work, this is usually the case--although even this requires some decisions to be made as routine work is not always routine and procedures do not always cover every consideration or circumstance.

How we make decisions is rarely covered when we are training. Somewhere along my journey becoming a trainer I realized I had been training Operators for ten years but did not understand how they made decisions. I would give them malfunction after malfunction on the simulator and hoped they figured it out. I never understood how I could help them interpret data, how not to get lost in all the data, and then how to use it to make a decision. It would be a new concept to understand how an Operator arrives at a satisfactory decision in a timely manner rather than unrealistically expect them to review all the options and pick the very best one without any deviation. Unfortunately, we do not always have time for to review all the options. I hope I have your interest.

## Types of Knowledge

Let's first look at what types of information we use to "know" and understand our processes. These terms will also help us continue our discussion further in this book.

### Explicit Knowledge

The type of knowledge we are most familiar with is explicit knowledge. This is the information we obtain from manuals, our routines, procedures, and other forms of documentation. It can be easily shared. It can be readily accessed and is usually easy to verbalize. Another term that may be used is declarative information. One simple example of declarative information: water boils at 100 degrees Celsius or 212 Fahrenheit.

Explicit knowledge is what is used in what we think of as analytical thinking. This involves the process of gathering relevant information and identifying key issues related to this information. This type of thinking also requires you to compare sets of data from different sources--identifying possible causes and effects and drawing appropriate conclusions from this data to arrive at possible solutions.

Knowledge alone can have its limitations. The information knowledge silo diagram is attempting to demonstrate that just memorizing information alone does not make knowledge useful. If we have had no opportunity to apply this knowledge it remains compartmentalized and unconnected. In our industry most parts of the process will have impacts on other parts of the process. The traditional forms of training and testing leave knowledge in this state.

The knowledge that is just memorized can be referred to as "inert knowledge" or "book smart". Tools to make knowledge useful in a dynamic process are discussed in the Chapter 5 on Training.

**Knowledge Silos**

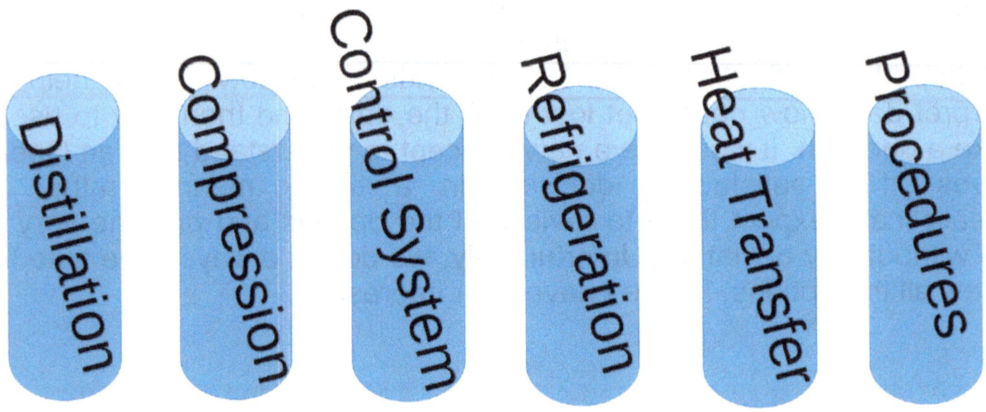


**Tacit Knowledge**

Another type of knowledge is tacit knowledge (job smarts). Tacit knowledge is harder to communicate and share. This is the skill of recognizing patterns based on your past experiences, practices you have developed, and your knowledge of what to key in on and what information is valuable. Tacit knowledge helps us to make connections between complex processes.

This type of knowledge is difficult to capture in a procedure. I just recently had a trainee experience this as he was reading through a procedure to start up a furnace. The procedure was split into sections that addressed the air, fuel, and feed all in separate sections in a linear fashion; however, it did not address that some of the steps happened concurrently, and that some of the steps on Page 3 happened with the steps on Page 7.

The author of the above procedure likely intended to address the air considerations all together to make it clear how the air system operated. Perhaps the author assumed laying the steps out in a sequenced approach may have taken away from that. Because the Operator was still developing his understanding of how these systems interrelated, he was behind on some of the steps while starting up the Furnace.

As this happened during simulator training there was little consequence. His trainers, including myself, could also own this confusion as we missed his lack of understanding.

A simulator is an excellent tool in developing tacit knowledge.

Another example of tacit knowledge could be when a Panel Operator is struggling with a complex problem, when someone walks up behind them, quickly glances at the graphic that may have up to 100 pieces of information on it and says, "This is your problem, and this is how you fix it". This skill will be discussed in more detail in the next section on decision-making with time limitations.

The diagram below shows how all the pieces of the same information seen in the explicit knowledge diagram are all connected and result in tacit knowledge or what we call "Experience".

# Process Mental Models

Your process mental model is the foundation of being a successful Troubleshooter. It is your mental representation of a system that allows you to describe, explain and predict the state of that system. The understanding of the components of a system--boiling points, partial pressures, chemical reactions, and process theory-- is more than just knowing information or facts about the process.

In the following diagram, we have examples of explicit knowledge in the boxes and examples of tacit knowledge in the cylinder shapes. This is not a comprehensive list and could include many more examples.

A mental model lets us imagine what might happen if we acted in various ways. A good mental model allows us to predict the effects of our actions (i.e., If I opened this valve 10%, what will it change and to what degree?). It will also help us diagnose what may be going wrong when a problem arises.

Our mental models are developed from a combination of our explicit knowledge and our experiences. They will also be influenced by our personal biases. Biases filter how we understand and see things. One key to keep in mind with our mental models of our processes is that our understanding of the process is only a model and not the real thing. When we are stuck, our mental model is likely wrong or inadequate.

**Example 1.** (Provided by Dr. Gary Klein)

We will use the control system for a refrigerator with a separate freezer as an example of how an inaccurate mental model can cause problems. You get your new refrigerator freezer home. Here are the temperature control instructions as provided on the door.

| | | |
|---|---|---|
| Normal Settings | C and 5 | 1. Set both controls |
| Colder Fresh Food | C and 6-7 | 2. Allow 24 hours to stabilize |
| Coldest Fresh Food | B and 8-9 | |
| Colder Freezer | D and 7-8 | |
| Warmer Fresh Food | C and 4-1 | |
| Off (fresh FD & FRZ) | and 0 | |

Even though it says to let it stabilize for 24 hours, you decide to turn up the freezer control to 8 so you can get the meat that you picked up while out shopping frozen quickly. The next day you notice that there is ice on the milk in the fridge. You turn the 'Fresh Food' setting to E. The day after that the ice cream is like concrete. Once more you adjust the 'Freezer' section to 5. Now the butter in the fridge is soft.

With only the controls to go on, your mental model of how the controls work may look like this.

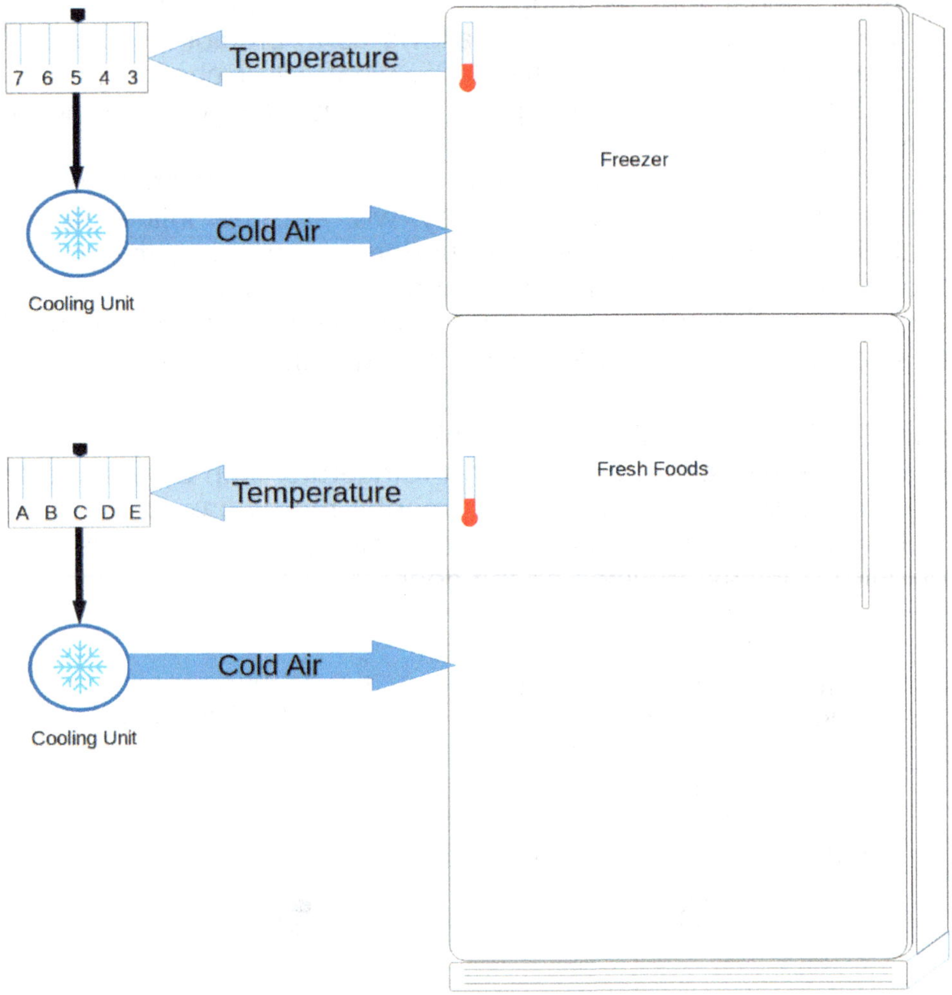


On an older refrigerator, this mental model would be incorrect. There is only one thermostat and one cooling refrigeration system. The freezer control knob determines when the cooling unit turns off and on, and the refrigerator control knob adjusts how much cold air is going to the freezer versus the refrigerator. Therefore, selecting a higher number will cause the cooling unit to run more and will drop the temperature in both the fridge and freezer. Selecting a letter closer to the beginning of the alphabet will cause the damper to direct more cold air to the fridge and away from the freezer.

This is a great example to show how easy it is to have misconceptions and the importance of understanding the nuances of control systems.[2]

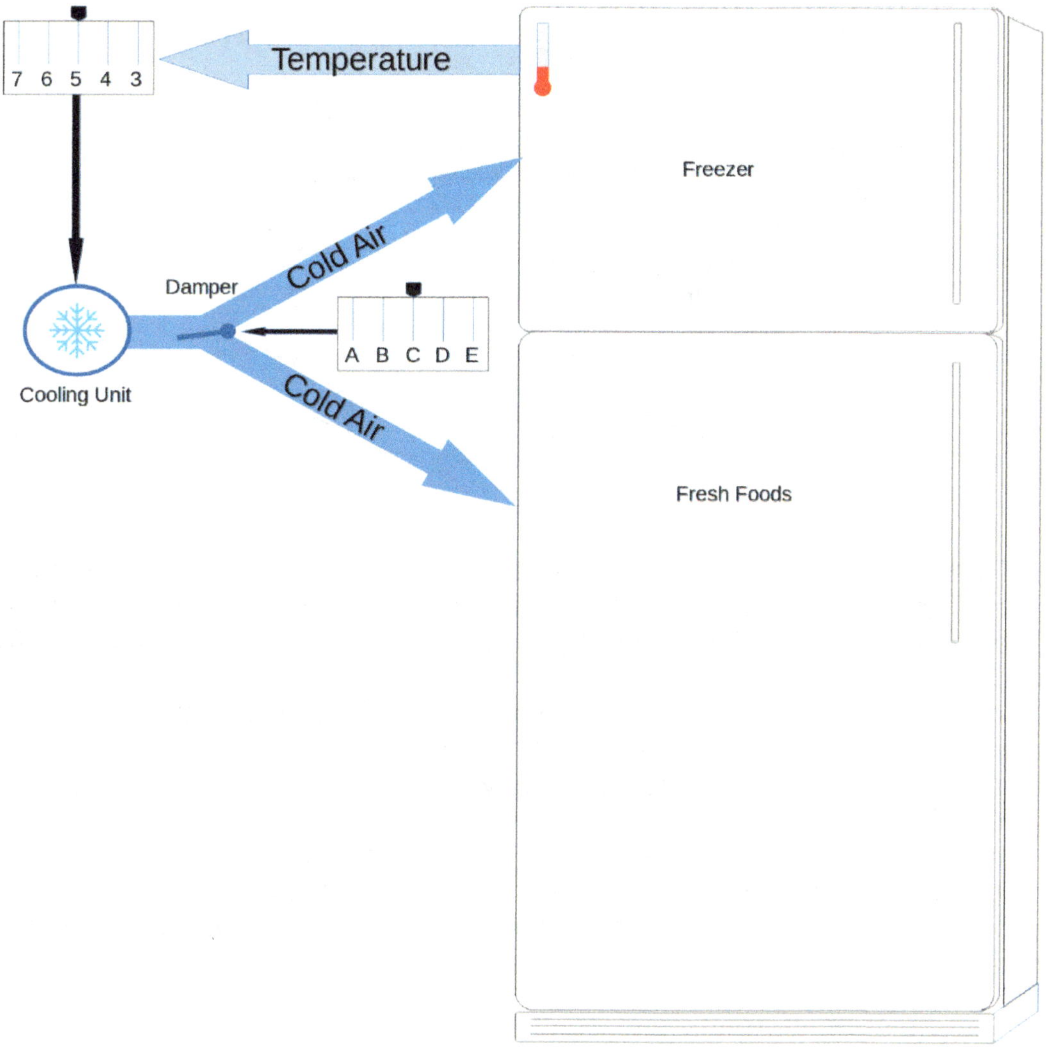


**Example 2**

This example is for a basic level controller on a vessel. If you look at LC101, the setpoint is 50, the process variable is 61 and the valve output is 100%.

Some assumptions that might be made:

- The valve is wide open
- The level transmitter is failing high
- The valve is plugged off

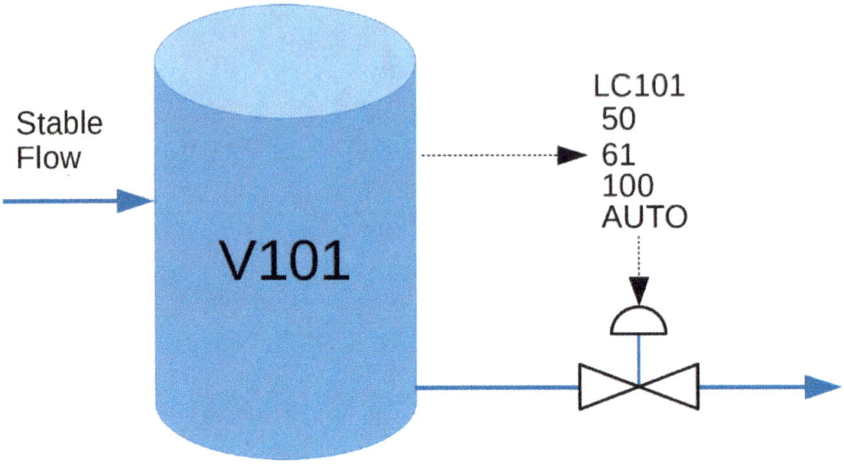


Traditionally, only the process variable was brought into the control room. The output shown on the graphic was only the request from the control system to the field. It could easily be assumed that the valve output indication is the position of the valve.

In this scenario, the valve is closed in the field. This could have been caused by the loss of electrical signal to the positioner, a failure of the positioner, or the loss of the air supply that drives the positioner and the valve.

In more modern systems, there is two-way communication with the positioner--one to give the output to the valve, and the other to indicate what the valve position is. The 0% indication in the diagram is feedback from the actual position of the valve.

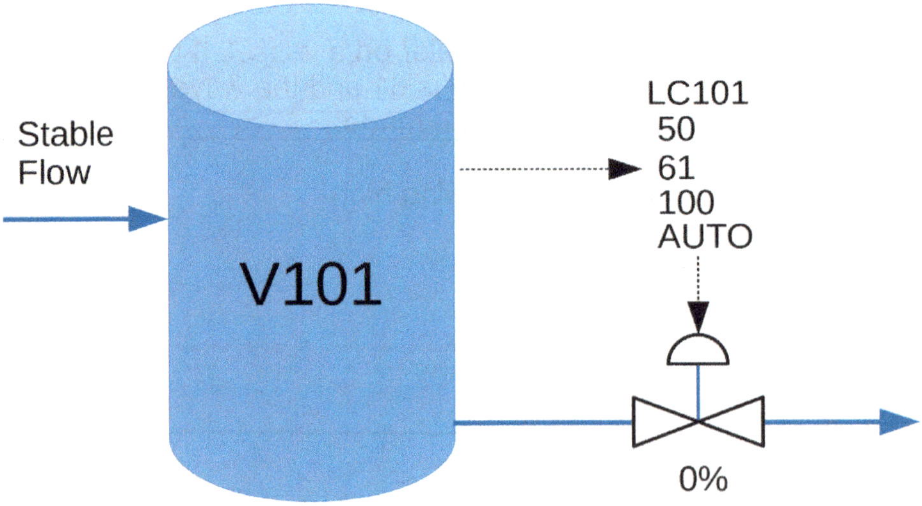

## Mental Model Matrix

In 2019 Dr. Gary Klein and Joseph Borders traveled from the hub of civilization in the eastern USA all the way to our location in western Canada. On this trip, they were researching the mental models of our Panel Operators using our CORYS high fidelity simulator. I must give the eight Operators credit for their willingness to go through complex scenarios while being watched by two cognitive psychologists. I developed three scenarios that the Operators had not trained on before and I rated two of the scenarios a 9 out of 10 for complexity. They were more successful at finding the problems than I was expecting, especially considering the distraction of spectators monitoring their every move.

Dr. Gary Klein discuss mental models in the paper "The Mental Model Matrix, Important aspects of mental models often get ignored", (published in Psychology Today[3]). The table below is taken from the article and summarizes the concepts from the article.

| | Positive (+) | Negative (-) |
|---|---|---|
| System | How the system works:<br>Parts, connections, causal Relationships, process control logic | How the system fails:<br>Breakdowns and limitations |
| Person | How to make the system work:<br>Detecting anomalies, appreciating the system's responsiveness, performing workarounds and adaptations | How users get confused:<br>The kinds of errors people are likely to make |

An understanding of how the system works is a common explanation of what a mental model is. What is not often considered is an understanding of how a system fails. What are the system's limitations and vulnerabilities to failing? Most of this understanding is what we gain from experience and will fall into the tacit knowledge category. It is not always shared in our manuals and procedures.

A key operating skill is to find a system's weakness. Many a engineer has left the control room deflated after an interaction with a highly suspicious Panel Operator after requesting they try something new with the process. A tip for engineers: before you go to the control room, think of all the problems that could happen and develop mitigations to manage them. Be clear of what you would like to achieve, and operations may have some suggestions. Some Operators will insist that this be well documented.

Operations should also have a willingness to try new things when all the concerns are addressed.

The next step after a failure is developing workarounds to stabilize the process, keeping on specification, and staying online when possible. Do not rule out the possibility that the best decision to make is to shutdown and isolate the facility (Safe Park).

One last aspect to consider about mental models is considering confusion. This could include the information presented in the graphics and also when communicating with others. Being very concise and precise with requests of the Field Operator can accelerate problem-solving.

## Dynamic Mental Networks

This concept expands on the Process Mental Model concept. It prepares us for the next skill that will assist us in making decisions in a dynamic environment. Each process move we make as operators might have several effects. This depends on the complexity of our processes and how much cross energy exchange, or rippling flow impacts other systems they may have on other systems.

Typically, when there is a problem, there will be more than one response required-- one action to correct the original problem and subsequent actions to mitigate the effects from the first problem.

Here are some examples:

**Time** - This addresses how long a response takes when a change is made. If the feed is increased to the front end of the process, then how long does it take to impact the back end? If you increase the heat to a converter, then how long before the reaction increases? If you increase the heat to a reboiler, then how long will it take to increase the heat in a distillation tower?

**Weather** - Cooling tower performance can be impacted by ambient temperature changes and humidity changes from storm fronts. Wind can blow the plume from a nearby cooling tower into the draft of another. Large rainstorms can cool distillation towers. Wind fluctuations can cause the draft to swing on fired equipment like boilers and heaters.

As an example, take a refrigeration system that supplies refrigerant to several systems and relies on other process systems' heat duty to condense the discharge vapors. This is an efficient system that needs a limited external source of cooling. In this system, a reboiler on a distillation tower could also act as a discharge condenser on the refrigeration system. The overhead condenser also uses the refrigerant from this system.

If the valve supplying heat to the reboiler fails closed, the distillation tower will cool down and go off-specification. With no vapor traffic going up the tower, the refrigerant requirements will drop off on the overhead condenser. Say there are another nine exchangers involved in the refrigeration system. The discharge and suction pressures on the refrigeration system will be affected. This will change the boiling points on the other exchangers using refrigerant and the condensing temperature of the refrigerant on the other exchangers that act as discharge condensers. These will then impact the systems they are associated with.

To make this a little more complex, the refrigeration compressor is driven by a steam turbine that exhausts into a low-pressure steam header. Load changes on the turbine could then impact both steam headers and any systems that rely on them. The steam headers could also be shared across the site with other units.

Describing this type of process interaction can be difficult enough. Trying to write a procedure to describe this can be very challenging if not impossible. Procedures are very linear in their format; therefore, they may not adequately describe complex, simultaneous interactions.

The ability to work through complex dynamic problems depends on our ability to interpret the cues and notice the patterns. Along with this, Operators must keep one eye looking forward and anticipating what will happen and the other looking back to validate what did happen.

Operators need to be looking for what should have happened, but did not, and anomalies that contradict their assumptions that may be telling them they are going in the wrong direction.

This is a constant questioning process that is driven by curiosity. Curiosity can be thought of as an emotive force that drives us to solve problems. If there is no emotion, there is no mental movement. Knowledge silos are created from a lack of curiosity about how everything is connected and works together.

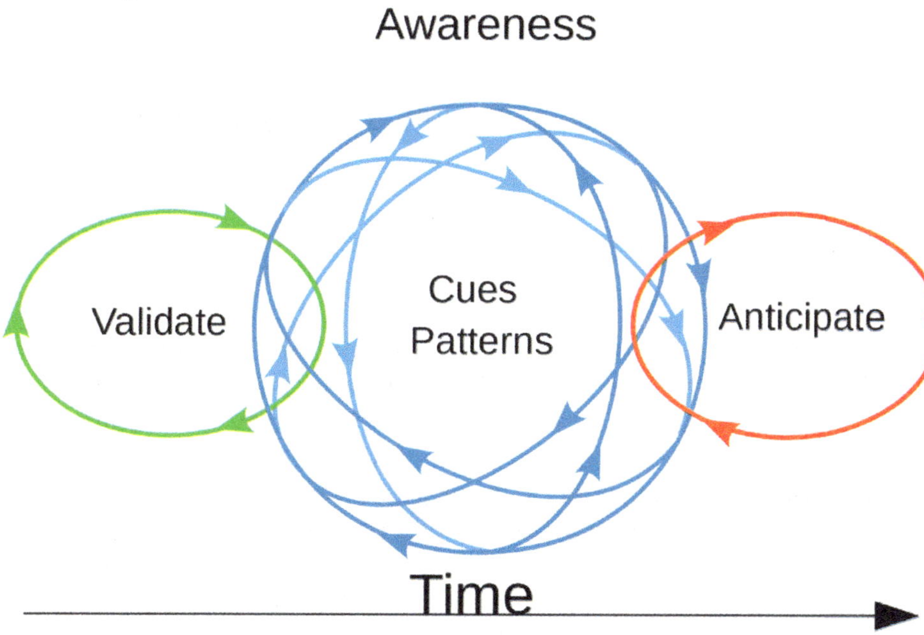

# Decision Making with Limited Time

Think back to a time when you had to handle a situation and there was some time pressure. Did you make a list of all the possibilities and then rationalize your way to a decision? Or were there some things you understood, and almost instantly you knew what to do? I found it helpful to list all the decision points I had to make in an event and then for each one identify how I arrived at that decision. Was it from experience, something you inferred from an experience or from a procedure? Perhaps from a process of elimination? If it was not X, then it must have been Y. What cued you into the decisions you made? This chapter is about how we make decisions.

Before reading further, think through a past event and document step by step how you arrived at each decision you made. You may not understand how you knew the answer to some of the problems--we will explore this in more detail later.

## Decision-Making Approaches

An experienced Operator, relying on tacit knowledge, may arrive at a conclusion almost immediately based on cues that triggered comparisons to past experiences. This can be very helpful, especially when there are time constraints.

Studies of decision-making in critical situations with time constraints have shown that approximately 90% of the time only one course of action is considered. There is not a comparison of options.[4]

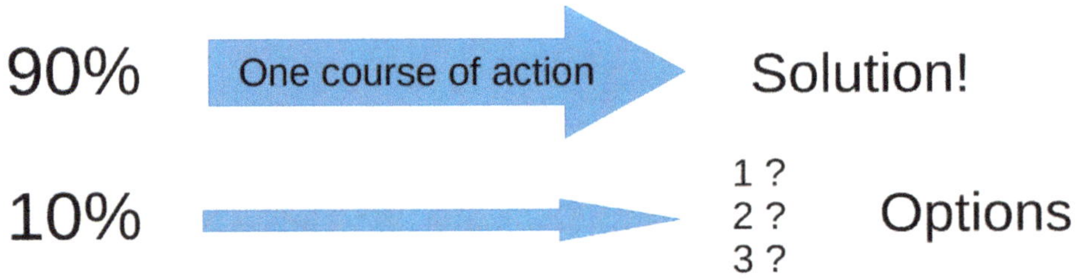


Some might think considering only one option is haphazard or even dangerous. In some circumstances it is hazardous to take the time to explore multiple options. For example, a response may be required before a constraint is exceeded, or trip logic is initiated. The situation could be isolating the process in response to a large leak to the atmosphere or a fire.

We will cover ways to develop the skills required to make these types of decisions in a later chapter. Of course, we prefer to take the time and examine all the risks when possible. We can also prepare beforehand and have a thorough understanding of the risks that will be used in these decisions.

One method to check the course of action is to perform a mental simulation (run it through your mind) before you implement an action. This would be taking the solution and asking what would happen if I ___________ (executed this action)? This would help you determine if you had to adjust your response or come up with a new solution. A thorough understanding of the process, which leads to a sophisticated mental model, is required to make this effective.

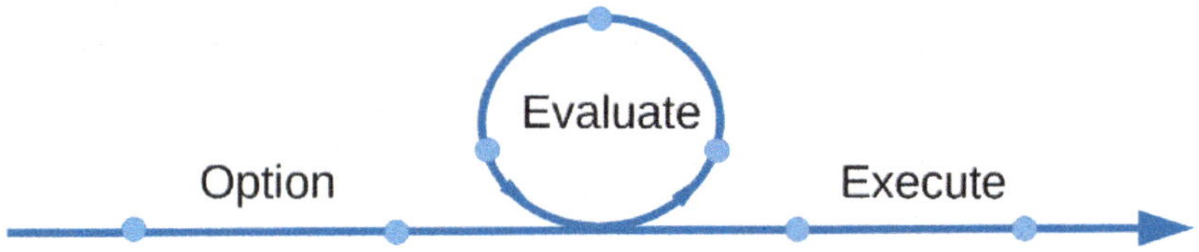


# Reality Check

Often situations are not the same and require some adjustments in the response. What worked last time may need some tweaking this time.

However, when there is not enough experience base, you are more likely to use a more analytical approach of comparing options. Unfortunately, comparing options in the absence of experience may not be very successful. It is the best that novices can do, but it is usually not good enough especially when the situation is deteriorating. Scenario-based training is an effective method to bridge that gap to gain experience.

## Pattern Recognition

Before we go further into looking at how we make decisions with limited time, let's look at one of the key principles that are involved in this.

**What are patterns?**

The Cambridge Dictionary defines patterns as "a particular way in which something is done, is organized or happens."

Our industry defines patterns as "a predictable sequence of events."

I believe we recognize patterns all the time without even being conscious of them. I will take that a step further to suggest that rational choice is about choosing between patterns.

Patterns have also been referred to as models to use to duplicate something (like a pattern used for making a dress). A more modern example would be the code used to make something on a 3D printer.

Ray Kurzweil's discusses patterns in his book "How to Create a Mind, The Secret of Human Thought Revealed"[5] For example, let's take the capital letter “A”. If you took the reasoning approach, you would look at it and think that there are two lines joining at the top and one horizontal across the middle. I would then compare it to the 26 letters of the alphabet and determine which letter it most likely is. You would not have to do this for every letter because you would recognize the pattern of the lines in the letter. The pattern of the letters then forms to make the patterns of words. You would not need to look at every letter in a word, let's say “Adam”, and break it down into the letters A-d-a-m, then compare it to the thousands of possible words and rationalize that the word is “Adam”. If we have done the work to develop this pattern, we will instantly know that it is the name of a person.

To take this to a technical level, we know that if a system has a trayed tower with an overhead condenser, a reflux drum, a reflux pump, a reboiler, a feed line, an overhead line, a bottoms product line, and several control valves, it is the pattern of a distillation system. I could see this configuration and instantly identify it.

Even this system is made of several patterns. The reflux pump has a driver, an impeller, a casing, a coupling, and seals. When these patterns are together, I know that when the switch is turned on the impeller will spin increasing pressure and cause the fluid to flow. I do not have to reason through this process every time I think of a pump. I will, however, need to reason through it the first time I am learning about how a pump works.

The patterns we hold in our minds are also models of what we believe the real world is about. Our minds are continually trying to make sense of our world and comparing it to the patterns that we hold. This could be called curiosity. I believe this to be a natural tendency unless it has been stifled by events in our lives. Although I believe this is changing, our traditional education system teaches us that we are holders of knowledge, which does not lend itself towards developing curiosity and seeking to understand how things relate to each other and form patterns.

Being able to recognize patterns is a key skill in making decisions when there are time limitations. David Eagleman discusses how experts have a less cognitive load in his book “Livewired, The Inside Story of the Ever-Changing Brain”.[6] This enables the experts to make faster and more accurate decisions.

Ray Kurzweil also discusses how humans have a weak ability to process logic but a very deep core capability of recognizing patterns. We must use our neocortex which functions as a large pattern recognizer and is not ideal for processing logic. This aligns with our understanding of humanity's pre-modern existence when we had to quickly interpret dangerous situations by using pattern recognition.

It has only been since our increasing use of technology that we are having to rely more on logical thinking. There has been such a focus on this that we seem to have forgotten about pattern recognition in our decision-making. Ironically it is scientific research that is rediscovering it.

Patterns are also thought to build off each other into a hierarchy of patterns. A core pattern may be used in combination with other patterns to build a larger pattern. This same core pattern may be used in several other larger patterns.

An example could be your understanding of how a control valve works. This same pattern or slightly altered version of it may be used in all the patterns we develop to understand the process systems that use a control valve. Concepts of the process will also work this way. Our concept of distillation will form a pattern and then can be used to understand the systems that use distillation. When we finally put together all the patterns of the systems, we then understand the process.

Research lead by Lisa Feldman Barrett, "How Emotions are Made, The Secret Life of the Brain"[7] has shown that our emotions are based on the patterns and concepts that we hold. Even to the point that they are predictive, which allows us to respond quickly to danger. This is then followed by feedback of what we see, hear, etc., and then we update our response. This may also have a downside as we may respond to an event and fail to update based on what we are perceiving (have you ever been in an argument, and you are arguing the same point?). The impacts of emotions will be discussed further in Chapter 4 on Stress.

## Recognition Primed Decisions (RPD)

For more than three decades, the cognitive psychology field of Naturalistic Decision Making (NDM) has been studying how people make decisions in complex real-world settings that can include dynamic and uncertain conditions. These decisions are for urgent situations with significant consequences for mistakes.

This research has taken place in the field with firefighters, military, health care, police and more recently in the petrochemical industry.

Dr. Gary Klein is one of the pioneers in this area of study. I would highly recommend reading Dr. Gary Klein's "Sources of Power: How People Make Decisions". This is the type of book that you want to take the time to reflect on the concepts and review some of the past decisions you have had to make. The ideas presented were a breakthrough for me in how I understood decision-making and how I responded to process upsets.

While I was a trainer, I covered the panel on occasion and would assist during upsets or start-ups. Rather than applying a narrow focus to problem-solving trying to step through each possibility, I learned to look for patterns with a sense of curiosity. This method of problem-solving is much less work and typically faster (not that I can explain how the subconscious mind works). Although we discuss this concept separately from analytical decision-making, we will likely use both concepts together to solve problems.

Refer to the diagram below which is based on a model Dr. Gary Klein used in his book "The Power of Intuition: How to Use Your Gut Feelings to Make Better Decisions at Work" [6]

- **Mental models** – Everything we perceive, and our interpretation of this information is filtered through our mental models (our beliefs and biases)
- **Biases** - Everyone has different mental models depending on their experiences that will influence their decision-making.
- **Situation** – Malfunction, process deviation, anomaly, contradiction
- **Physiological Responses** - This is how stress can affect us. The Window of Clarity in Chapter 4 covers this in more detail.
- **Cues**
  - **Panel**, Visual: information from process graphics (alarms, process variables, trends, video), audible alarms
  - **Field,** Visual: leaks, gauges, tactile: temperature, vibration: audible: leaks, flow changes, PSVs lifting, smell
- **Patterns** – This is where we notice the relationships of all the information we have observed. We use our Mental Models that we form from past experiences and training.
- **Compare to Normal** – Because we work in a man-made environment it is more difficult to recognize patterns unless we are familiar with what the normal conditions are. Trends are an excellent tool for this.
- **Action Scripts** – These are the possible responses to the situation.
- **Mental Simulation** –

  Based on the chosen action, imagine what would happen if I ______

Recognition Primed Decision

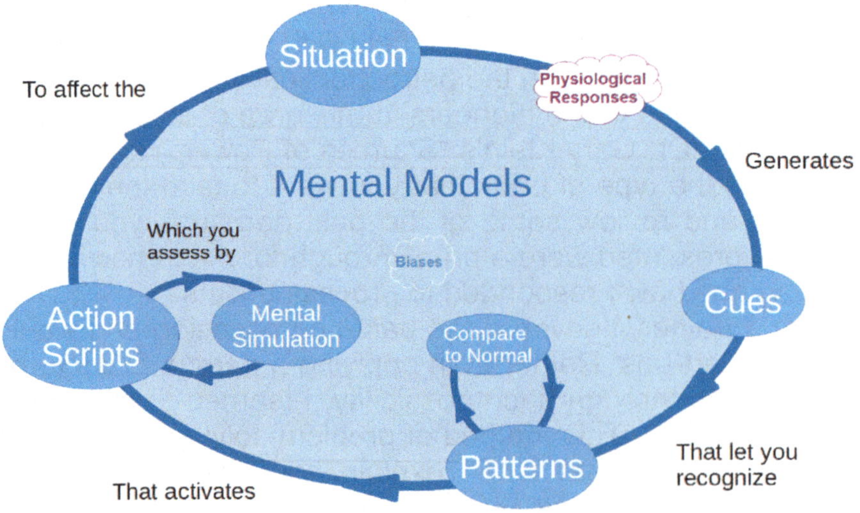


Adapted from "The Power of Intuition" pg. 26, by Dr. Gary A Klein

## Problem-Solving Mindset

It can be overwhelming and easy to get lost in the data. It can also be difficult to explain how an experienced Operator can sift through all the available information to get to the source of a problem. As pattern recognition is mostly a subconscious process there is no step-by-step approach. There are concepts that we can use to help us develop a mindset that can help us work through problems.

Below is a diagram of how to resolve the problem of getting lost in the details. It is a combination of utilizing pattern recognition and analytical thinking. Consider this another dimension to the previous RPD model. The mind is an extremely sophisticated biological system that scientists are just beginning to understand. It will take several different models to try and portray how to use it most effectively.

In our complex industrial processes, there is rarely just one event. There can be several secondary impacts that may require attention or intervention. If we only zero into the initial problem we may miss the secondary problem. Or worse we may zero into a secondary problem first before we notice what the initial event was. We can become consumed with trying to stabilize a controller that is just responding to the initial problem while the initial problem is escalating. To prevent tunnel vision, we need to come up for air occasionally and think about what else is going on.

In the model below, the aim is to be open-minded to all possibilities, be curious, and continually scan for new patterns. When we pick up on a cue we zoom in and focus on the details. We must remember to zoom out to an overview mode occasionally to see what else is going on and to explore other possible solutions. Focusing in on the details is great for analyzing problems but does not help to find alternatives.

One other reason we need to zoom out occasionally is to see if the initial problem may be causing another part of the process to reach a trip state or go off-specification. This can be termed “Tunneled In”. From this perspective, we may miss the big picture or the rest of the secondary impacts.

We may also fail to see that we are focused on something that was impacted by the initial problem and never see the initial cause as it spirals out of control. Once we understand what the cause of the problem is and move into executing the solution, we will spend more time in the detail and occasionally look at the big picture. (Refer to diagram on next page)

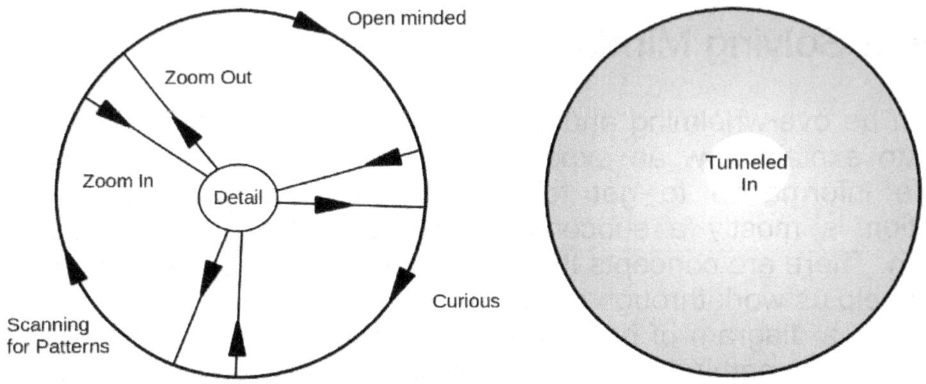


**Troubleshooting decision considerations**

The diagram on page 30 is an adaptation of the RPD model to reflect decision-making in our industry. Although this may look like a series of steps, experienced Operators will not follow these steps in any order. The marvel of our subconscious mind is that it may work through all these considerations almost immediately. Experienced Troubleshooters often arrive at their conclusions very quickly and may not be consciously aware of all these considerations.

I am breaking this example down into a high level of detail to help novice Operators start to build a framework that they can use to solve problems.

This might seem like common sense to some. (I believe the term common sense can be used when we cannot explain why we know something. This term is sometimes used in a belittling fashion which allows us to deflect from our lack of understanding of our own thought process.)

This section intends to help the novice Operator to reflect on all the considerations that may happen during a troubleshooting event. The best way to integrate these skills is to review past incidents and practice going through scenarios.

In the first stage in troubleshooting, we are typically trying to quickly assess the situation and minimize any impacts. The amount of time we have will depend on the process. If you determine that you have more time, you can move into a troubleshooting process that involves more data gathering and validating before a response is required. This is covered in Chapter 3 and referred to as Stage 2 in the process below.

Problem-solving is a dynamic process. Operators receive information in many forms, they draw meaning from this information and then they interact with the process. The model below demonstrates this with three main loops. The arrows in each loop are there to demonstrate that these three actions must be considered continually during an upset, as conditions can change rapidly requiring a modified or different response.

## Decision Process for Upsets with Time Constraints

Refer to diagram on next page

**Monitor** –During steady-state conditions or during an upset, it is important to continually look for feedback that the process is responding as expected.

**Evaluate** –This involves looking for cues and patterns that are familiar to past incidents or similar incidents and making choices on which direction to proceed.

**Stabilize** –Action will be required if the process is about to exceed material specifications or process specifications. This could be placing controllers in manual or executing an emergency procedure and safe parking the process. This may be the first step required even before there is time to complete a thorough evaluation.

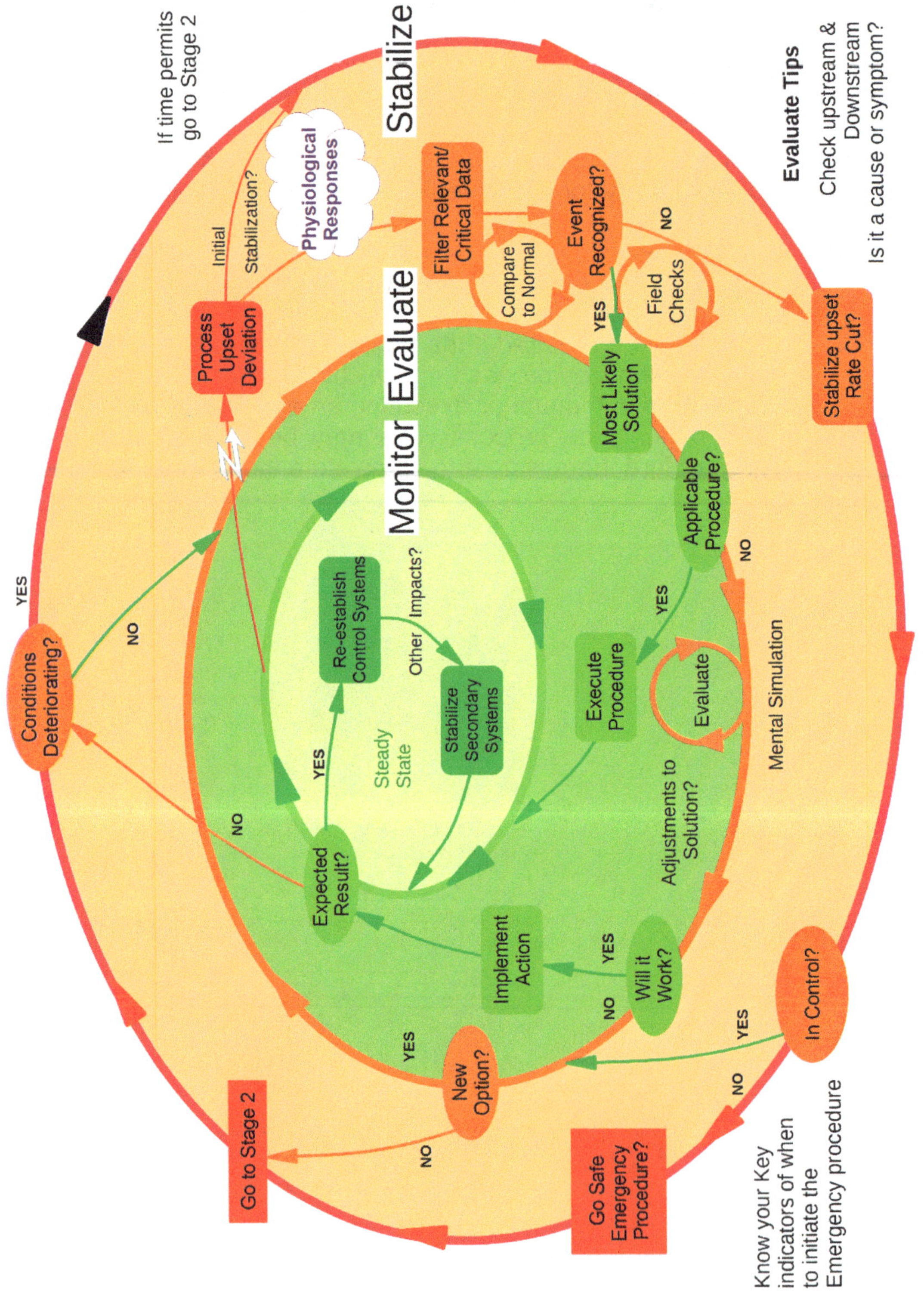
If time permits go to Stage 2
Stabilize
Evaluate
Monitor
Evaluate Tips
Check upstream & Downstream
Is it a cause or symptom?
Process Upset Deviation
Initial Stabilization?
Physiological Responses
Filter Relevant/ Critical Data
Event Recognized?
Compare to Normal
NO
YES
Field Checks
Stabilize upset Rate Cut?
Most Likely Solution
Applicable Procedure?
NO
YES
Execute Procedure
Evaluate
Mental Simulation
Adjustments to Solution?
Will it Work?
YES
NO
Implement Action
Expected Result?
YES
NO
Re-establish Control Systems
Other Impacts?
Stabilize Secondary Systems
Steady State
Conditions Deteriorating?
YES
NO
New Option?
YES
NO
Go to Stage 2
In Control?
YES
NO
Go Safe Emergency Procedure?
Know your Key indicators of when to initiate the Emergency procedure

| | |
|---|---|
| Process Upset Deviation | This is the protagonist in the story where we are leaving steady state operation. We become aware of this from an alarm, noticing changes from a graphic or from the field operator. You may also have Event Prediction software available. You first response may be to stabilize the process. This could be placing a controller in manual and returning it to its previous output if the controller was fluctuating wildly. There could be any number of process changes that could place the system in a safer position, and this depends entirely on the process. We will cover some of these in a latter chapter. |
| Physiological Responses | These are the physical responses that stress can induce. Typically, there is some form of surprise during a large event that may impede our thought process, even for a short period of time. There are other responses and techniques to mitigate them that will be covered in the Window of Clarity chapter. |
| Filter Relevant/ Critical Data | One of the more challenging steps of troubleshooting is filtering the data and picking up on the cues. We have access to thousands of points of information and are now able to bring more information into one graphic. Which has its advantages but comes with challenges. Alarms can help us to zero in on the problem, but they can also mislead us. The alarm could be a result of a response not directly related to the major problem that needs to be handled. We can also experience alarm flood where we have so many alarms all of our attention could be used just reading alarms. Trends may also be helpful at this stage.<br>The question is how do you key on the critical information without long deliberation? We typically do not have time for lengthy deliberation if our goal is to keep our process running and prevent damage to equipment or personnel. Knowing which pieces of information to monitor and which to ignore is another type of tacit knowledge. To filter data you have to become familiar with where it is located. This means taking the time and effort to know your graphics. Review them until you understand what every point is and what the key parameters are. What matters the most on each of the graphics? Having a sense of what normal is for the current conditions can also be very helpful. This will help to key in on the trouble areas quickly — the anomalies. You can't spot an anomaly if you don't know what is typical. |
| Event Recognized? | This can happen quickly if you recognize patterns from previous experiences or training  If you recognize this situation and already have a pretty good idea on how to resolve it you are halfway there. You do not have to take the time to compare possibilities. If you spot any anomalies that make you uncertain, then you need to take some time to examine more carefully what is going on. Some field checks may also be required. |

| | |
|---|---|
| Most Likely Solution | Based on past experiences you may already have a solution for the problem. |
| Applicable Procedure? | If there is a procedure already developed for this situation, implement it and then monitor to insure it was effective. There is a chance it may need to be modified for this situation. |
| Evaluate<br>Mental Simulation | If there is not an applicable procedure, then ask yourself what would happen if you implemented your solution. Completing a mental simulation can verify if your action is likely to be successful or if needs any adjustments. This requires an accurate mental model and the ability to forecast/predict what will happen if an action is implemented (upstream/downstream implications). |
| Will it Work? | If you believe it will work based on your mental simulation, you can move onto implementing the action. If not, you will need to come up with a new option. |
| Implement Action | If your Mental Simulation looks promising, then execute this action and watch for the result. |
| Expected Result? | Monitor the situation to insure it is going as planned. If so proceed to Re-establish Control Systems. If not move to Conditions Deteriorating. |
| Re-establish Control Systems | If you were successful, then place the controllers back to their normal mode. A new problem could develop if you leave a controller in manual and are not monitoring it. It can take several hours to stabilize the process back to normal after an upset. |
| Stabilize Secondary Systems | You may also have to stabilize secondary systems that were impacted from the original problem. Avoid getting lost in these secondary systems during an upset. At times they do require attention but if they become the main focus the initial problem will get ignored and the situation will deteriorate.<br>These considerations are the normal steps when you successfully troubleshoot a situation that was straightforward to resolve. However, don't expect to be so lucky. Often there will be some trial and error or even a total change in course required. The following sections will attempt to address these situations. |
| Conditions Deteriorating? | After you have implemented your actions and you do not get the expected result, you will have to evaluate if the conditions are deteriorating. Then decide if you have to focus on stabilizing the process or if you can re-evaluate the problem and come up with a new plan of action. |
| Stabilize upset Rate Cut? | If the problem can't be easily identified (Event not Recognized) then efforts will have to be made to stabilize the situation as best as possible. This may mean placing controllers in manual, adding an additional pump or reducing |

| | |
|---|---|
| | feed rates to that system or the whole plant. This may be required until the outside Operator has time to respond to give you additional information or to allow time for a more methodical review of the problem. This typically takes extra resources while the primary panel Operator manages the upset. It can be challenging to respond to the consequences of an upset while at the same time going into a deeper evaluation of the problem. Involve another panel Operator, Supervisor or Process engineer where applicable. |
| Go Safe Emergency Procedure? | If the process cannot be kept in control or on specification, a partial or complete shutdown may be required. This can be a difficult decision and it is best to give this a lot of thought beforehand at what conditions you would have to do this. Determine which process variables you would use and their limits beforehand. This choice can go against all the efforts and training that help us to keep the plant running. The first several minutes of your response to the emergency procedures should be memorized and reviewed frequently. Your mental abilities will be taxed enough without having to dig deep in the memory bank about what steps are required to secure the plant. This can lead to mental overload and freezing up mentally. You should review the main steps of the Emergency procedures at least once every few weeks and more when you are first learning them. After you have executed an emergency procedure, review the procedure and ensure it was completed and continue to monitor the process for secondary problems. Secondary problems include leaks, de-pressuring /sub-cooling, over-pressuring, high temperatures, passing valves, high or low liquid levels, and any other abnormal conditions. |
| Go to Stage 2 | This is a stage where the cause is not obvious, and you must consider other options. A few options may be easily evaluated with the above thought process, or you may have to go to what we will call a Stage 2 approach that usually involves more resources. We will cover this further. |

After reviewing all of this you may be saying "What happened to rationalizing through the problem and picking the best option?" That can happen at any of these decision points. The event-recognized step has the most potential for having to work through more than one possibility. If you do not have any background with this type of situation, you may have to consider more than one option and pick the best possibility. If the situation is escalating, you may have to move to the stabilize mode until you can develop a plan.

## Comparing approaches

Now that we understand what a Recognition Primed Decision is, we can compare it to rational decision making and where we may use each. During an actual process upset that can be quite dynamic, we may use a mixture of these two approaches as we may have several decisions to make.

I will borrow a table from Dr. Gary Klein's "Sources of Power" that compares the two strategies.

| Task Conditions | Recognition Primed Decisions | Rational Choice Strategy |
|---|---|---|
| Greater time pressure | More likely | |
| Higher experience level | More likely | |
| Dynamic conditions | More likely | |
| ill-defined goals | More likely | |
| Need for justification | | More likely |
| Conflict resolution | | More likely |
| Optimization | | More likely |
| Greater computational complexity | | More likely |

When one course of action is decided on:

- **Time pressure** - The consequences of waiting may be greater than making a decision that is close and adjusting it as needed.
- **Experience** - More likely to have seen the situation or something similar.
- **Dynamic** - The decision made becomes useless by the time it has been implemented because the process has already changed.
- **ill-defined** - Ambiguity makes it hard to come up with the criteria required when the goals are not clear. This should not be too common in our industry.

When a comparative evaluation is used:

- **Justification** - You may have to explain your reasoning to leadership why you made the choice you did.
- **Conflict resolution** - It is important to evaluate different priorities before deciding.
- **Optimization** - It is worth a thorough evaluation to reduce the risk before pushing the process further.
- **Computational complexity** - There are times when using math will get the best result. For the most part, these types of decisions have been made for us when the facility was designed. We simply follow the design limitations.

## Case Study: Converter Off-Specification

**Nitrogen that was not completely displaced from the process**

When I read Dr. Gary Klein's "Sources of Power" I used this incident to validate the concepts for myself.

There are two parallel caustic towers available to allow for cleaning of the towers without stopping production. One tower was cleaned, and the oxygen was removed using nitrogen. It was then pressure purged with process gas to remove the nitrogen. This tower was slowly brought on-line, and then the back-up tower was taken off-line. The purpose of the caustic tower is to remove carbon dioxide from the cracked gas stream.

Not all the nitrogen was removed, and it reduced the purity of the hydrogen stream that was used in an acetylene converter further downstream. The reaction was C2H2 (acetylene) + H2 (hydrogen) = C2H4 (ethylene) in the converter and its purpose is to convert acetylene to ethylene. This led to acetylene breaking through the converter which would result in the product in the pipeline going off specification.

As the converter was going off specification, I more than doubled the flow of hydrogen, and we then requested to receive a backup supply of hydrogen from another facility on site. This happened quickly since we had called them earlier as a precaution. The converter quickly came back onto specification, and the excess hydrogen on the outlet stayed quite low even though it had twice the amount of hydrogen required. I chose to reduce the amount of hydrogen flowing into the reactor even though the hydrogen concentration leaving the converter was quite low.

To say I was anxious about the direction I chose is an understatement. This was not a procedural step and was not documented in the manual. If I was wrong, I would have had to flare all the production and have the plant feed reduced to 2/3 within 5 minutes to allow the flaring to burn clean. If I left it in this state, the temperatures could have continued to rise until the safety system automatically isolated the hydrogen supply and flared the converter effluent automatically. The C2H2 in the product did increase to the pipeline but did not exceed the limit of 5 ppm.

So how did I know to take this approach? I did not have time to consider several options or calculate how much energy would be generated from hydrogen combining with other molecules in the converter. Neither did I have time to consult with our process engineers to come up with an explanation.

I based this decision on an experience we had bringing the reactor on-line during a start-up. As the converter effluent is flared until the outlet is on specification, we thought we would add more hydrogen flow than normal to speed this process up. The amount was the same as we used for upset conditions. This did not work, and we then had to reduce the hydrogen flow and the inlet temperature to get the converter on specification.

The excess hydrogen was increasing the bed temperatures, and the catalyst which normally preferred the $C2H2 + H2$ reaction was also causing the $C2H4 + H2$ reaction. This was taking away some of the hydrogen that was required for the C2H2 conversion and caused the outlet to be off specification. We had to reduce the hydrogen flow and inlet temperatures to bring the converter on specification.

For me, this was an example of how a Recognition Primed Decision works. I recognized a pattern from a similar incident, slightly modified it for the circumstances, and successfully applied it.

# CHAPTER 2
# Problem Solving

## Insights

One of the keys to solving a problem is gaining insights. This is the recognition part of the Recognition Primed Decision model. There is no magic formula for this although there are keys on how they can be encouraged. Below are some concepts that are taken from Dr. Gary Klein's book "Seeing What Others Don't - The Remarkable Ways We Gain Insights". He discusses three ways that insights are developed. Since we are discussing insights, come up with your own ways to encourage insights as well, and I would be glad to hear about them.

The first variety of insights comes from creating desperation. This is the state in which we are willing to consider almost anything. Our thought process is challenged, and all our efforts are geared toward finding a solution to this problem. Our beliefs and assumptions are challenged until they are reworked and help us to find a solution. Keep in mind that the problem is not unsolvable; it is our limited understanding and beliefs that are preventing us from solving it.

Desperation is what keeps us up at night and what might lead us to wake up at 3:00 AM with the solution. This is what may lead us to 'testing the waters' -- a technique where we try something but only a small amount, so we can see if it makes a change but not such a large one that it would cause a huge problem if we were wrong. In our industry, this may be increasing or decreasing an output on a valve slightly to see what happens. We do not understand the problem yet, so we start making some calculated moves to see if the process is responding as expected.

The second variety of insights comes from connections. This is trying to understand how ideas or concepts fit together and looking for patterns from the information we are gathering. How does an upset in one system relate to another system? How does a feed composition change impact the functioning of the whole system? We may receive alarms in two different systems. We must ask how they could be related or if they are independent problems. If they do relate, which one was the cause, and which was the effect?

The third variety of insights addresses contradictions. Am I seeing something that does not make sense compared to the other information I have? Does the increase in flow indication to a system correlate with the decrease in flow out of the system? We use the term mass balance to describe the balance between the inputs and outputs of a system.

There is typically some error in every transmitter. Check historical flows for what the normal offset in the mass balance the system typically has. Applications can be built that will calculate the mass balance for you and display it on a graphic.

Checking the mass balance occasionally for offset and then having the transmitters corrected will be very beneficial during an upset. This could involve having the transmitter's sensing lines cleared or recalibrating the transmitter.

A transmitter will be calibrated based on a certain pressure, temperature, and composition of the stream. Weight flow tags can be created that will take these factors into consideration to provide a more accurate reading.

As you can see, it is easy to delve off into technical issues when discussing insights. The way information is presented in the graphics can be very helpful in noticing anomalies.

I will add a fourth key to gaining insights and that is through analogies. Has there been another situation that may provide a solution or give an insight for a new solution? The solution may need to be adjusted to fit your current situation.

Keep in mind that if you have proceeded down a path of correcting a problem and contradictions are popping up, then it is time to reassess your strategy and it may be time for new insights.

## Curiosity

Curiosity is a mental state that keeps asking why? During training curiosity helps you to connect the dots. It is the motivator that keeps you looking for insights that will help you solve problems. If there is no curiosity, there will be no mental movement. This concept is occasionally mentioned except the conversation does not go very far, mostly because we are not that proficient at instilling curiosity in someone. Rather we are very good at stifling it. It is directly tied to the 'culture' or atmosphere of our work environment. We will not be successful Process Operators without a well-developed sense of curiosity.

Using a curious mindset every day will slowly build your knowledge, explicit and tacit. It will help to connect the dots and develop a complex neural network to utilize when solving problems. When your curiosity has led you to a new understanding, take a minute to feel the satisfaction of these successes no matter how small. Encourage others to develop their curiosity by asking leading questions without giving away the answer. Develop a culture where it is OK to ask questions and explore new possibilities. If someone misunderstands something or is taking the process in the wrong direction, try to understand what their thought process was. This is discussed further in Chapter 5 - Training, under Learning from Mistakes.

## Assumptions

Most of us have either experienced or know of a situation in which a course of action was implemented but the situation continued to deteriorate because the Operator never questioned the original assumptions. Our industry has too many examples of this. It can even happen when an experienced Operator suggests a solution and the others on the team may be reluctant to question it.

If you are stuck, then retrace your thought process and look where you may have made an assumption that could be false or incomplete. Refer to the section on a problem-solving mindset and keep a broader perspective, don't become stuck with tunnel vision. It can be difficult to know that you have made an assumption. If you are stuck on a problem, and it is not returning in the direction you had intended, it is time to re-assess your beliefs about what is going on. You may also have to consider that you do not understand the system as well as you thought, and you may need new information or more expertise.

I know of a few situations where the Operators (I include myself in one of these) were unable to solve the problem, and it took someone else to walk into the situation to solve it. Sometimes groupthink can set in, and everyone buys into the same assumptions. This is more likely when someone in a position of authority suggests, and everyone is uncomfortable questioning it. If it takes leaving the room and walking back into the room with a fresh perspective, then try that.

We also may experience what is called confirmation bias. This is when we filter the information to support what we believe is true. We only look at the indicators or information that prove our theories. Be watchful when you start to dismiss anomalies that do not agree with your theory. One may not be unusual, but when you have 2 or 3, it is time to question your beliefs.

It is also helpful to understand how we perceive information. Our eyes are connected to the thalamus, and the thalamus is connected to the visual cortex. There is 6 times as much information communicating to the thalamus from the visual cortex as there is from the eyes to the thalamus. You will want to read that line again. Most of what we see is an internal model of reality that is continually updated from what we see. Watch this

YouTube video to expand on this further. - "The Brain with David Eagleman - What Is Reality - BBC Documentary 2016" At 28:40 he addresses this topic.[9]

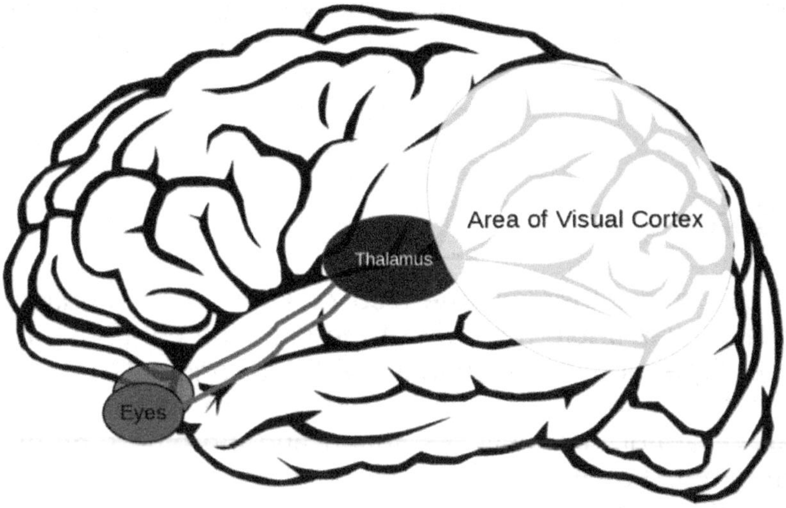


There is 6 times as much information passing from the visual cortex to the thalamus as from the eyes to the thalamus

Brain image by Gordon Johnson from Pixabay,[10] additional elements by author

Think of the implications of this from the point of view of gathering information from graphics. Is what our mind sees your internal image of what is there, or is it the updated image? Maintaining your focus and curiosity will ensure you are updating your mental image.

**Sometimes the wrong question is being asked.**

On the next page, is an example of a basic trayed tower with a steam-driven reboiler. We are only focusing on the bottom section of the tower. The first diagram is the system in steady-state operation.

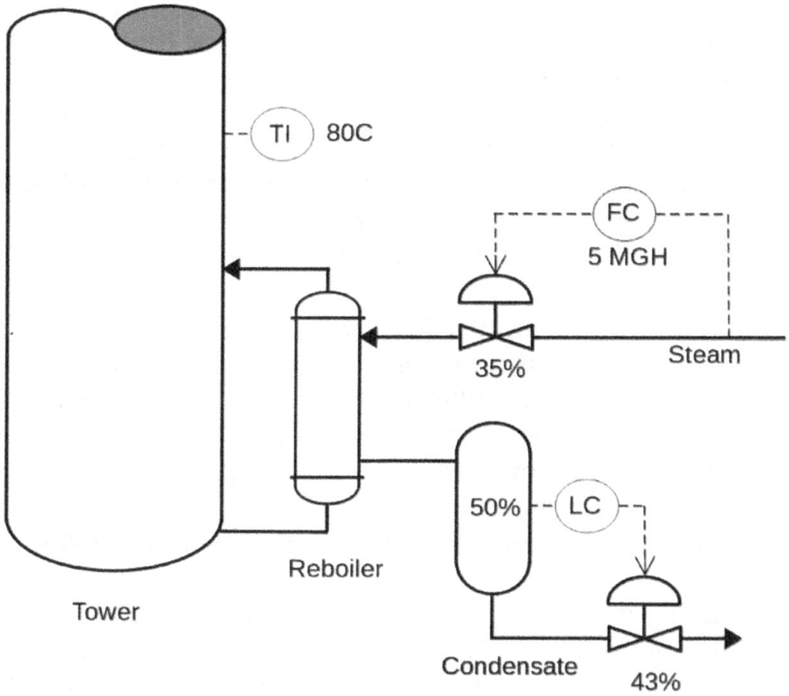


The next diagram is the same system with the tower cooling off from 80 deg. to 70 deg., the reboiler flow is indicating 0 MGH (megagrams/hour), and the valve is 100% open. The level control valve on the condensate pot is closed while the level is still at 50%. It is assumed the rest of the system is in steady-state operation.

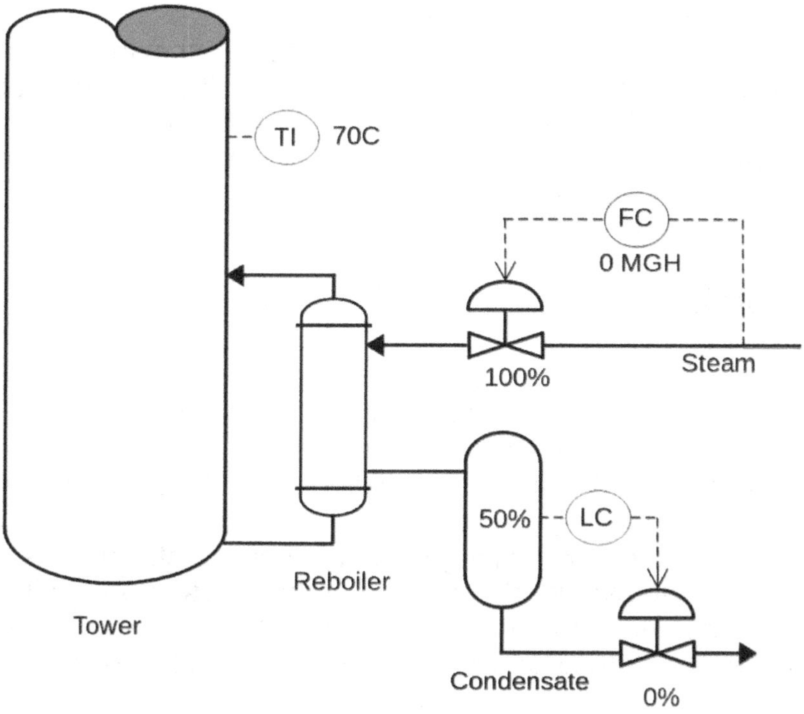

Initially you may think the problem is the lack of steam flow. Perhaps the valve failed closed (the condensate level is higher than indicating), or the steam pressure dropped.

But the real question could be "What caused the loss of heat exchange?" In this example, there were non-condensables in the shell side of the reboiler that would not allow the steam to enter the reboiler. The non-condensables could be process gases from a leak or nitrogen that came in with the steam from another part of the process.

**When troubleshooting does not follow the process**

We have tried to systematically break down all the steps in a troubleshooting process. Hopefully, we have that "ah-ha" moment and figure it out quickly. Frequently it is not that easy, and the process is much more fluid and requires some innovation. It is difficult to exactly model the mental process that someone will use to solve a problem. Ask someone who solved something quickly, and they may have difficulty describing how they "knew".

Troubleshooting is dynamic and requires constant evaluation of whether your responses are effective. Sometimes Operators must make decisions without all the information and make the best decision we can with that limited information. If we were to wait for all the facts to be evaluated and compared, it may be too late. There may be times when we will have to decide with the best information we have. We will be doing this with many things taxing our mental processes. That's another reason to develop thorough mental models of the process and to use scenarios to increase your tacit knowledge.

Procedures can be written for our responses to known upsets. The more consistently the upsets follow the same pattern, the more successful the procedures will be. The rest of the time we will have to use our mental models and tacit knowledge to determine the best course of action.

## Critical Thinking

A definition of critical thinking is the following: the process of actively and skillfully conceptualizing, applying, analysing, synthesizing, and evaluating information to reach an answer or conclusion.

In the Psychology Today article called "The Second Wave of Critical Thinking"[11], Dr. Gary Klein writes:

"The concept of critical thinking is clearly important. There is widespread acceptance that we would do better if there was more critical thinking. Society would benefit. Students would do better in schools. People would be more successful in their personal and professional lives.

"We generally view critical thinking as making judgments based on the systematic analysis of evidence of some problem or topic, rather than relying on impulses, opinions, and emotions. That's the standard approach, which is the first wave of conceptualizing critical thinking."

**The Second Wave of Critical Thinking**

Dr. Gary Klein then goes on to discuss the Second Wave of Critical Thinking. In this state of critical thinking, we are looking for patterns and cues to develop our perception of what we are seeing. We can see how systems connect and interact with each other. Typically, critical thinking is seen as a cause-and-effect relationship--if this happens then it will result in this action. In our complex systems, one action may result in several other actions, and those actions may result in more and maybe even going back and impacting the first action that started the sequence. This can make it difficult to understand what started what. It is even more difficult to write a procedure to describe this.

Dr. Gary Klein continues:

"The second wave encourages us to use our judgment about what's happening in situations and what should be happening. Instead of blindly accepting the interpretations of others and passively following orders, we can do more to actively make sense of events.

"We can be alert to weak signals that others aren't noticing. We can become curious about coincidences and possible connections. We can be sensitive to events that were supposed to happen but didn't. We can be more mindful. Curiosity appears to be a common denominator for all of these aspects of the second wave of critical thinking.

"One aspect of critical thinking is a shift in mindset, from a procedural mindset to an investigative mindset. The procedural mindset is that all we need to do is to follow the procedures. Certainly, in most jobs, we need to learn the procedures. But sometimes the rules and procedures don't apply or would lead to poor outcomes. Procedures are necessary, but they aren't sufficient. We need to learn the procedures but shouldn't get trapped by them. We also need to look around."

**Preparation**

Critical thinking is best done before an event happens. If you are relying on critical thinking during an event, your mental capacity can easily be overloaded, and you will be more reactive to the situation. It is very helpful and necessary to logically think through the process to understand the relationship of the different systems i.e. What happens when this control valve opens more? What happens downstream and what happens upstream of the valve? What other systems does it impact (and there may be several)? Even this level of critical thinking is about pattern recognition. There is more on this topic in Chapter 5 - Training.

These exercises will develop mental models of the process and enable us to recognize patterns that will assist us in the abilities discussed in the second wave of critical thinking. In the second wave of critical thinking, we are comparing patterns as possibilities to solve our problems. This is a more complex form of critical thinking that involves many more variables.

## Procedures

There are different lines of thought about procedures. Some believe everything will go well if the procedures are followed exactly--that there will be no problems.

In one of my roles, I wrote 20% of the procedures for a new petrochemical facility. We had a thorough review process with three engineers that were involved with the design of the facility, three operations specialists that had also been involved in the design since the start, and five operations technical leaders. I would like to say we were 75% accurate to how the plant ran, but I may be biased.

I can certainly say that I was challenged every time we started up, as there were dynamics we did not anticipate. As with most new facilities, there were a couple of false starts. We were fortunate to have a simulator to allow our Panel Operators some experience before they operated the actual plant. This also offered us the opportunity to refine the procedures as well.

Even with these challenges, I believe procedures are essential. Our processes are too complex to memorize every step, and not having to think through every step every time reduces the workload. Procedures are great training tools for novices and help to compile experiences or learnings from incidents. There are critical tasks that require specific sequences of events that if not followed could lead to lost production, hazardous releases, or worse.

Procedures are a great tool for documenting what steps are required for known emergency or large process upset situations. The steps required can be discussed and evaluated when there is no time pressure. These situations can include power failures, instrument air loss, the loss of any equipment that requires the shutdown of the whole process, feed losses, responses to off-specification events or even large rate reductions.

It is also helpful to determine ahead of time what the key indicators are required for you to determine which emergency procedure to initiate. You can have the procedure memorized and practiced, but if you are uncertain whether you should execute it, you may stall.

Procedures can impose consistency and help crews to coordinate across shifts. This is beneficial for activities that take more than one shift and helps the relief crew know where to take over. There are also complex tasks like taking equipment out of service for regeneration or to remove fouling that can take coordination of the Panel Operator and the Field Operator. If they have not completed this task together before, then a procedure can ensure they are both approaching it the same way.

Procedures, practices, and permits are a must for preparing equipment for maintenance. Preparing equipment is a critical task for operations to minimize the risk to maintenance personnel. This takes a coordinated effort between the two groups to ensure the equipment is safe to work on and safely returned to service. There is more on returning equipment to service in the Start-up section in Chapter 9 - Extras.

For all the strengths of procedures, they can fall short. They are great in well-ordered situations but become brittle in complex situations that can have many interactions. Every procedure is written with a set of assumptions that usually are not articulated (i.e., steady state, all equipment available and operational, full rates, etc.)

It is difficult to write a procedure that is in a linear format to capture a multi-variable process where several things could be impacting each other at the same time. A plant start-up is like this. There can be two or more Panel Operators working on different systems that will affect the other systems. A procedure is a good resource for a start-up; however, if you are completely relying on the steps and not interacting with the conditions that present themselves, you will have a challenging start-up.

It is difficult, if not impossible, to anticipate every possible situation that could happen. A few years ago, a company came and toured my plant's simulator facility. The visiting company had attempted to write a procedure for every possible problem that could happen. This was not working as well as they had hoped, and they decided to change their approach to improve Operator performance via simulation-based training. Even if you could write a procedure for every possible problem, which I believe would take hundreds or more procedures, how would you find the right one in time to respond?

## Case study: Amine Upset

**Amine Contactor foaming in a Sour Natural Gas Processing Facility**

An amine contactor is a trayed tower with the sour natural gas flowing up from the bottom and amine is pumped in above the top tray (see diagram on next page). The DP (differential pressure) across the tower is measured to indicate the presence of foam. Foam restricts the gas flow, therefore, increasing the DP. The amine removes the hydrogen sulfide (sour) from the natural gas.

Foaming issues had been experienced in the amine contactor for several weeks. There had been several theories proposed to explain this phenomenon but every attempt to solve it had been unsuccessful. The only way the foaming could be kept in check was by injecting an anti-foam solution which alleviated the foaming by reducing the strength of the film of the bubbles.

During one night shift, there were two events when the high DP alarm came in at 17 KPa (kilopascals, or 2.5 psi). An injection of anti-foam returned the DP to the normal pressure of about 11 KPa each time. At around 2:20 AM, the high DP alarm rang in again. The Chief Operator gave the amine system another anti-foam injection and waited for the normal two to three minutes it takes for the DP to drop. The DP continued to increase to 21 KPa and another injection of antifoam was tried.

As the DP continued to climb the amine temperature was increased by three degrees. This can help as the viscosity of the amine is reduced with the temperature increase and this reduces the flow resistance. A third injection of antifoam was also made at this time.

As the DP increased to 28 KPa the Chief Operator decided to reduce the gas flow through the contactor by having the outside Operator shut down a compressor. When the gas flow started decreasing the amine flow rate was also reduced. The steps had no effect as the DP surpassed 32 KPa.

At this point, the sweet gas knockout drum downstream of the contactor started filling up rapidly as liquid carried over. The gas flow out of the contactor was shut in to prevent further carry over.

By this time the contactor had gone sour, so the flare valve was opened allowing the off-spec gas to flow to the flare system until sweet again. This took approximately three-quarters of an hour. An attempt was made at bringing the contactor back online but the moment the gas flow was initiated the DP rose quickly, so it was shut in again.

It was now shift-change so the day shift Operators took over, leaving the night shift Operators in a state of depressed shock, wondering what they could have done differently to prevent this catastrophe.

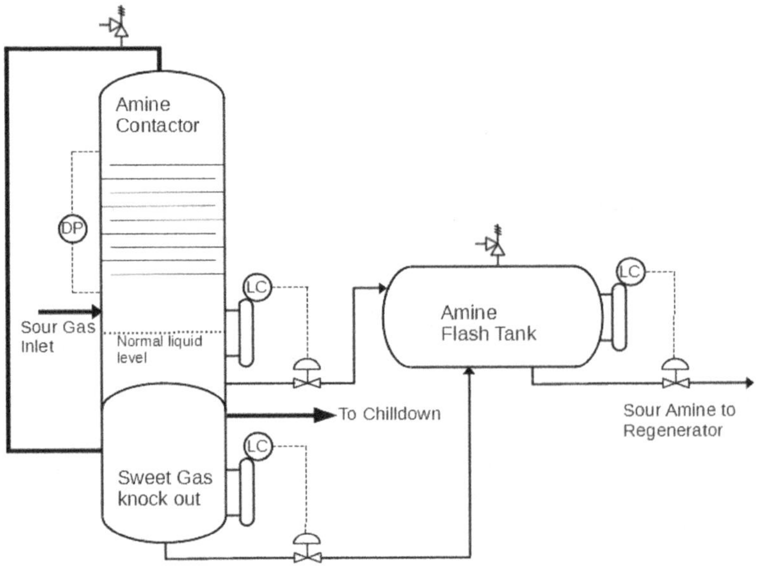

**Post-mortem**

It was found that the severe foaming occurring in the amine contactor had been caused by a higher than the indicated liquid level in the bottom of the contactor. The level measurement relayed to the control room had been gradually deviating over time from the actual level due to a pinhole in the displacer float used in the level sensor. Due to a coating in the gauge glass, the actual level was difficult to ascertain. The accuracy of a gauge glass also declined during a foaming event.

In hindsight, since all the other steps that were taken by the Operators failed to rectify the situation the question could be posed: "Why didn't they consider liquid level as a possible factor?"

As usual, this was not as obvious as it seems for the following reasons:

Until this time, the anti-foam injection had always been successful and sometimes was only needed once or twice per shift, and occasionally, not at all.

When a pinhole leak occurs in a displacer float level sensor, the deviation from the sensed level to the actual level can occur so gradually over a long period of time (weeks or even months) that there are no obvious signs as it is happening. An interesting footnote: due to this event, a decision was made to switch from the displacer float to a 'radar' level sensor which is more accurate and reliable for this service.

Lowering the level in the amine contactor is not a decision made lightly since too low of a level can cause high-pressure gas to flow to the amine flash tank. This downstream tank operates at 1/10 of the pressure of the contactor. This would cause the PSV to lift and result in a major flaring event.

Here's the kicker and the most difficult one to face— the "human factor". This event occurred in the middle of a night shift with only two Operators in the plant (a company cost- saving measure). Both Operators had over 60 years of experience between them but neither one had ever been through a foaming event they couldn't control. When it's the middle of the night on your third night shift, and you haven't been sleeping well during the day, your brain circuits sometimes just refuse to do their job. This is when you fall back on experience. If this event hasn't occurred in your career, you could be in for a rough ride! And here's the paradox: as you get older, a good sleep tends to become more and more difficult. So even though your age may have given you valuable experience, there will eventually be such a sleep deficiency that the trade-off between that experience and the ability of your sleep-deprived brain to function adequately will be too great to overcome.

**Contributed by Clary Michael**

# CHAPTER 3

# Troubleshooting With Less Time Constraints

This certainly is the preferred method to troubleshooting and problem-solving. There is less stress and more time to evaluate different possibilities. We can take the time to thoroughly evaluate any actions before implementing them. The process does not always give us this much time but when it does it is prudent to use it.

This process involves gathering information and possibly the right people. Having the involvement of the experts in the problem area is quite helpful. This may also involve sending the experts off to another room while you deal with the upset.

The first step is to start documenting information. The best way to do this is to use a worksheet like the one below. A worksheet gives a more structured approach; it is less likely that information will get lost from one shift to another and efforts will not be repeated. There have been many varieties of these created. Develop what seems appropriate for your situation. I have included some examples as well.

## Data Gathering

**Problem Description** - Give a short description of what the problem is.

**What are you observing?** – What has changed from normal operation? Provide a summary of what is going wrong. Be careful not to make assumptions at this stage or it may steer the whole investigation in one direction.

| | |
|---|---|
| Problem Description | Flow capacity is limited from the bottom of Quench Water Stripper tower, T101 |
| What are you observing? | Control valve opened to 100% and the pump discharge pressure dropped |
| When did the problem start? | Progressively getting worse over the last few hours |

**Process Variables** – Document all the controllers and indicators involved in that system. This can include valve position indicators, analyzers, and switches. Note if any component is off normal. Ask yourself what the condition of that component is. Could it be the cause, or is it a symptom of the actual cause of the upset? If a transmitter is not used for a controller or advanced computer control, then it must be a symptom and cannot be a cause. A diagram or copy of a console graphic can be a helpful reference. Instrumentation tends to have the highest failure rate so it can be worthwhile validating it first. This step also helps to ensure all the data is collected to validate possible causes.

| Tag | Descriptor | Abnormal? | Describe anomaly | Cause or an effect? | Possible failure mode |
|---|---|---|---|---|---|
| TI10051 | T101 feed temp. | No | | | |
| FI10010 | T101 feed flow | No | | | |
| FC10010 | Steam flow to T101 | No | | | |
| TI10152 | T101 overhead temp. | No | | | |
| LC10100 | T101 bottoms level control | Yes | High level | Effect | |
| FC10110 | T101 bottoms flow control | Yes | Low flow | Possible cause | Valve positioner may be failing |

Another helpful technique is to place before and after diagrams beside each other to compare the difference. This is easily done if your historian tool has graphics already built in to view past data. You could also use a screen capture of your control system graphics and either draw in the normal values or use Microsoft Paint (or similar) to edit in the values. Or, create your diagram and save them for future use as I have done here. There may already be diagrams that could be used from the training manuals.

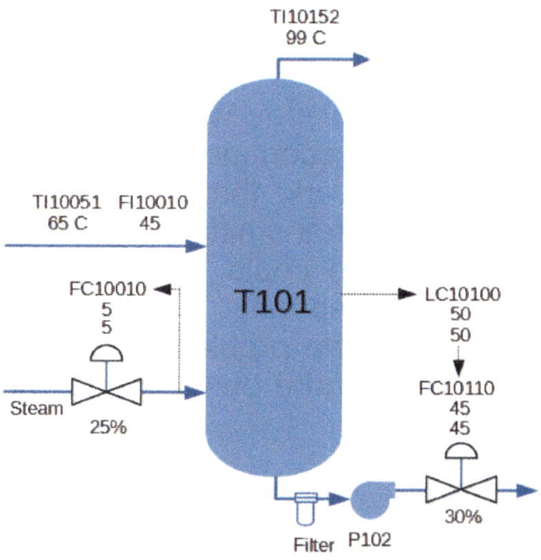


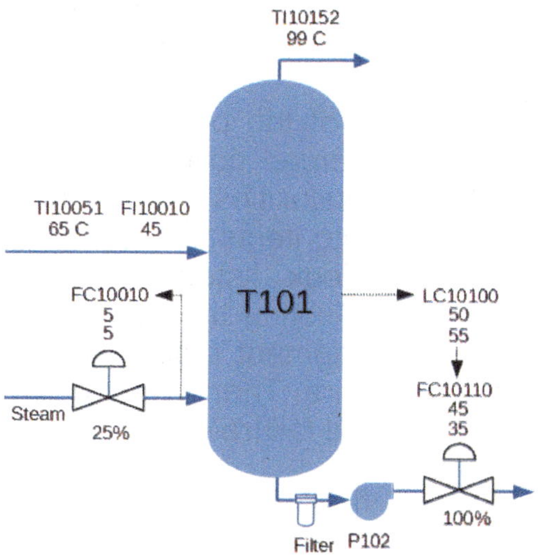

Then, list any possible mechanical causes.

**Possible Causes** – The next step is to brainstorm any other possible causes for the problem. It can be helpful to document all the possibilities before proofing them. This keeps an open mind and open discussion resulting in more possibilities being considered and the possibility of the right one being among them. The answer might be from the quiet voice who is hesitant to share their thoughts. Of course, this process depends on how much time you have. Sometimes this process may have to be abbreviated due to process demands.

**Does it explain the observations?** – Go through a cause-and-effect discussion to see if the possibility holds up. Explain how it could be true.

**Historical or background information** – Has something recently been changed or adjusted to this equipment or system? Is there a history of past problems?

**Actions required to confirm cause** – How can it be proved or disproved if the cause is a possibility? Sometimes it can be easier to prove that a possible cause is not true to narrow the options. This may be researching existing information, contacting a vendor, or completing testing. It is best to start with the easy confirmations that only involve field checks or stroking a valve, etc., then progress to checks that involve taking equipment out of service and then inspecting or dismantling.

**Results of confirmation** - Documenting each step is important to prevent data from being lost or held by someone that is away the following week when the effort continues.

**How will the cause be repaired?** – Determine who and how the problem will be resolved. Setting timeliness can be helpful.

After most of the possibilities are gathered you may wonder where to start. My esteemed colleague, Randy Saunders, recommends eliminating any of the possibilities that do not require extra work, through reasoning. Then move onto items you can visually check like transmitters, trends, and field checks (look, listen and feel). The next step could be testing your theories by adjusting the process to see how it responds. The last step is taking the equipment offline to inspect it (unless the problem demands it, like a seal failure. Then you are left with trying to figure out what may have caused the seal failure).

| | | | |
|---|---|---|---|
| **Possible Cause** | Valve positioner may be failing | Pump impeller is worn or fouled | Pump suction filter is plugging |
| **Does it explain the observations?** | Yes, the flow would be reduced if a valve is partly open | Yes, pump is unable to meet flow capacity | Yes, the pump may be cavitating from a drop in suction pressure. The water is near the boiling point. |
| **History or background information** | Positioner was upgraded to Smart positioner 2 years ago | Pump was overhauled last year. No sign of impeller wear or fouling | There was a flow disturbance the previous evening caused by an upset in the upstream system |
| **Actions required to confirm the cause** | Check valve output in the field<br>Check output from Smart positioner | Isolate pump and inspect impeller | Check pump discharge pressure gauge<br>Isolate and inspect filter |
| **Results of confirmation** | Output indicates 100% open in the field and from Smart positioner | | Pump discharge pressure was low<br>The filter canister was partially plugged |
| **How will the cause be repaired?** | | | Clean the filter canister |
| **Was the repair successful?** | | | Yes |

Other questions:

- There could be more than one cause i.e., an exchanger could have some degree of fouling accentuated by a change in approach temperature. There have been many incidents where the problem escalates because a secondary warning device did not function. Digital on/off switches used to be quite common as a secondary warning. These were often attached to a float that would sometimes stick or fill with the process fluid.
- What is not happening but should be?
- Check upstream and downstream of the problem to help determine what the problem is and to verify that your prognosis is accurate. If you believe the problem to be a level transmitter to have failed low, then the system downstream should be low in inventory. Use the flow transmitters and level transmitters to validate this.

Below are some additional items to be considered

| Additional checks | Follow up required | Details |
|---|---|---|
| Have the equipment material specifications been exceeded? | | |
| Did a PSV (pressure safety valve) lift during the event that needs to be inspected? Rupture disk? | | |
| Was the equipment depressured and chilled below the minimum allowable temperature from the process vaporizing? | | |
| Are there any other secondary problems that resulted from or were highlighted because of the incident? | | |
| Did any leaks develop as a result of the incident? | | |

Set up electronic folders to store this information for future reference. Create subfolders for each unit to make it is easy to locate in the future. Pictures or any other documentation can be stored here. A link from the operations web page can also be helpful.

It is best to have access to the process history, equipment manuals/files, logic descriptions, P&IDs, control documentation, and if possible, the experts in the area where the problem is. This may take more than one discipline until the problem is narrowed down. This could be a combination of any of the following: Process Operator, Process Engineer, Mechanical Engineer, Instrumentation Technician, Electrician, Mill Wright, Rotating Equipment Specialist, Fired Equipment Specialist. (Note: These positions may have different titles depending on your location and company).

# Premortem

This is a technique developed by Dr. Gary Klein with one of his early companies. A premortem[12] is the hypothetical opposite of a post-mortem. A post-mortem is a medical term used when someone has died to determine the cause of death. A premortem is imaging into the future that something has failed and determining why it has failed. This could be thought of as troubleshooting before the problem occurs.

One use for a pre-mortem in our industry is when there is an experimental change in how the process is operated or a physical change to the process from the addition of new equipment. Traditionally, when a change is being planned, a risk analysis occurs. The risk analysis consists of a group of experts breaking down each step and determining what the risks are, the severity of risk outcomes, and how to mitigate the risks. If mitigations exist to reduce all risks to an acceptable level, the change could be implemented. A negative part of risk analysis, however, can be the participants' hesitancy to be negative about a project.

In the pre-mortem style, the aim is to simulate looking into the future after the change has been implemented and that it has failed. Everyone is given a few minutes to predict why it could have failed. Then starting with the project lead everyone is asked to provide one reason from their list. Everyone states a different reason until all of them have been documented.

In this process, there is less reluctance to highlight potential problems that could come up. As Operators, we are typically pretty good at this type of exercise. We get a little anxious every time there is something that could affect the operation of the facility. This can be a little disconcerting for the engineers who are trying to promote the change. You can always enlist the Operators to come up with ways to mitigate their concerns. Having the Operators involved that will be implementing the change will also sensitize them to early signs of trouble during the implementation.

# CHAPTER 4
# Managing Stress

"Sometimes moves have to be good enough, you have other priorities"
David Strobhar, Author "Human Factors in Process Plant Operation"

## Window of Clarity

I believe stress is one of the least acknowledged issues Operators must face, and it can have a large impact on decision-making and performance. This certainly was one of the challenges I faced when I became an Operator. Our schools typically train us to memorize information, and that there is one right answer, and you have to be very precise to get full marks. The mindset that this develops is not helpful in a real-world dynamic event.

Stress can be defined as what you think you can accomplish and what you believe is required of you in a demanding situation. The lack of ability to manage stress has traditionally been seen as a weakness--either you can "cut it" or you can't. My experience has been that individuals that had little tolerance for other's stress were themselves working at a higher level of stress and tended to aggravate all those around them. This typically shows up as dominating the situation with little tolerance for other's ideas--usually accompanied by a "There is only one way to manage this situation--my way". Fortunately, we are beginning to understand stress and its impact on us, and there are tools and techniques to help us address it.

The airline industry can provide us with an understanding of the demands that can be placed on decision-makers. Capt. Chesley "Sully" Sullenberger successfully executed an emergency water landing of US Airways Flight 1549 in the Hudson River off Manhattan, New York City, on January 15, 2009, after a flock of Canada geese shutdown both the engines.

In a CBS 60 Minutes interview, he was quoted as saying that the moments before the emergency water landing were "the worst sickening, pit-of-your-stomach, falling-through-the-floor feeling" that he had ever experienced." "The physiological reaction I had to this was strong. I had to force myself to use my training and force calm on the situation". This interview can be found on YouTube (Capt. Sully Sullenberger and USA Flight 1549 Crew - 60 minu)[13]. There was also a movie made of the event called "Sully" that is worth watching. It stars Tom Hanks and was directed by Clint Eastwood.

You may not think Operators are making decisions that have life and death consequences daily until a disaster happens. As well as safety systems may be designed, not every situation can be anticipated. Operators need to try to anticipate events and respond accordingly.

The diagram below attempts to show the effects of stress and fatigue on performance. It demonstrates the relationship between stimulation and performance. The stimulation is all the demands that are placed on an Operator from alarms to process upsets, phone calls, radio calls, etc. If there is no stimulation, we become complacent and disinterested and then our performance suffers. As the stimulation increases, we become engaged, and our performance increases. There becomes a point when we have too much stimulation, we become overwhelmed, and our performance drops.

There can be many variables that may determine where this optimal point is, and I have tried to show how fatigue can reduce the level of optimal performance. Because of this, the optimal performance is likely more of a wave that fluctuates during the day as the demand on us changes and our level of fatigue changes. This would suggest that if we are at optimal performance all the time without breaks, our fatigue will start to undermine our performance. Fatigue will be discussed further at the end of this chapter.

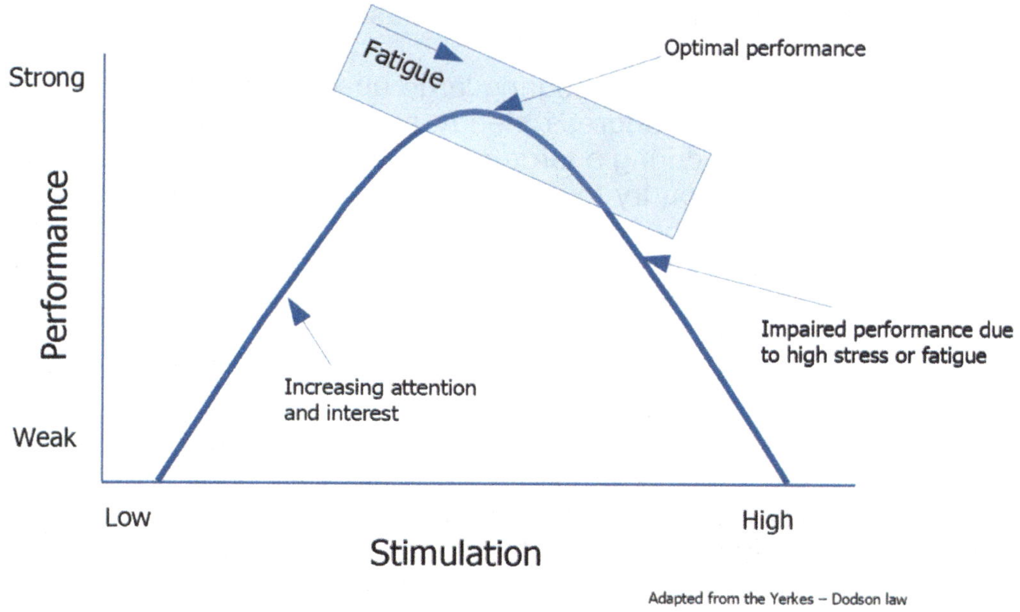

There will always be some uncertainty. In some circumstances, the selected decision will have to be acted on and then evaluated for its effectiveness. It is preferred to take the time to evaluate all the options and pick the best one. However, there will be circumstances when there will not be enough time. In these situations, we can use a mental simulation with your preferred course of action to validate it and to become aware of any unwanted consequences.

Interruptions are another form of stress and can have a large impact on our performance. Our thought process can be lost due to other distractions, alarms, or people. A conscious effort will be required to refocus. A Beville engineering newsletter, (Spring 2016 - David Strobhar) mentioned a study by the Australian military (Shayne Loft, Andreas Sadler, Janelle Braithwaite, and Samuel Huf, "The Chronic Detrimental Impact of Interruptions in a Simulated Submarine Track Management Task", Human Factors 57.8 (2015): 1417-1426) shows that in a command-control task (what console Operators do), interruptions can degrade subsequent situation awareness and task performance for up to 40 seconds. It can be so easy to be side-tracked and forget you have a valve in manual, a drier you were supposed to put into a regeneration, or an alarm that came in.

Here are some strategies that can help manage interruptions:

- Hand off or make a note of secondary issues that do not need your immediate attention. I used to keep a sticky notepad beside me to put notes right on the monitor for critical reminders and used a notepad for tasks and reminders to enter data in the electronic log we used.
- You will continually need to evaluate the priority of other issues that arise. Try to determine if it is a symptom or a cause. Could the new issue that arose be the result of another issue?
- Utilize alarm horn suppression during large upsets when available. Alarm rationalizations typically address steady state operation, and there can still be an alarm flood during a large event like a plant outage. Constantly hearing alarms while you try to isolate the process can be distracting and will lead to ignoring all the alarms until the task of isolating the process is complete. This does not mean you have to suppress the emergency or urgent alarms. A risk assessment will need to be completed to develop the strategy used.
- Shelve or disable nuisance alarms. How you accomplish this will depend on your operating system. If you have a transmitter that is constantly ringing in and out, it may become a larger hazard than if it is inactive. I have seen Panel Operators be conditioned to ignore critical alarms because of faulty transmitters or limit switches constantly ringing in and out. They had to make a choice to either run the process or constantly respond to a repeating alarm. This is a critical situation that should be addressed as soon as possible.

- Manage or remove distractions in the control room. Everyone is curious and wants to be helpful during process upsets. Some may be required to be in the control room. Others could be redirected to a different room to work through problems where they will not distract the Panel Operators. If you are one of the individuals in the control room and are required to have a conversation with someone else, take it away from the Panel Operator. Their stress level is high enough, and you will have a detrimental effect on their performance. Yelling or screaming will also have a large impact on the Panel Operator's performance, a mostly negative one. However, pointing out a critical piece of information that is missed could be helpful.

**In the Hot Seat**

You have read all the procedures, and you come across a situation where the procedure does not help you. How do you manage this? Your heart may be racing, your throat dry, palms sweaty, a knot in your stomach, and you may not even be able to get your mind working.

You are not the first or last Operator in a critical situation that this has happened to. It happened to me in my first major process upset until someone gave me a nudge. It is hard to say how long I would have been there without the nudge. This is an involuntary response by our nervous system.

If you are stuck, accept that you do not know the answer but that you do have the ability to find the answer. If you start to be concerned about what people are thinking or what they will say if you fail, then just breathe, let the thoughts pass, and refocus on the task.

**Unthinkable**

Amanda Ripley is an investigative reporter for Times magazine. She wrote the book "The Unthinkable, Who Survives When Disaster Strikes - And Why". This is directly applicable to our industry and offers very helpful strategies if you travel in airplanes, spend time in large buildings like hotels and convention centers. This book is about boosting your survival odds. We always hope we will never be involved in a large process disaster and have an event of the likes that the Chemical Safety Board covers. I believe no matter how much effort we put into prevention, there will always still be some risk--less risk, but it is still there.

While all accidents are preventable, you cannot prevent all accidents. In hindsight, it's avoidable, but that requires knowledge you didn't have a priori.

One of my motivations for writing this book is because of how unprepared I felt making some critical decisions as a Panel Operator and as a Shift Leader. I hope the following approaches will help with your critical decision-making.

One of the areas we tend not to like to talk about are the physiological responses to stress. These are the ways that our minds and body can be impacted from critical situations. This does not have to be a life-threatening situation. Operators sometimes make decisions that can have huge financial impacts and can become very public, from incident investigations and the very noticeable flaring.

## Physiological responses:

- **Freeze up** - Brain reverts to primitive response which overrides higher functions. Freezing can be a helpful response if you notice a wild animal and you do not want them to see you. Most people do not normally panic. They are more likely to freeze and minimize the severity of the event.[14]
- **The perception of time may change** - This can be either slower or faster.
- **Tunnel vision/fixation** - In this state it is challenging to see the big picture. This a very common response that can easily lead to an upset escalating. If the root cause of a problem can be determined and the main control systems of a process understood, an upset can be stabilized with less effort. A good example of fixation is in 1972 when an Eastern Airlines jet crashed outside the Miami airport because the landing gear light was burned out. They were so focused on the landing gear they lost altitude and crashed. The rest of the plane was functioning correctly.
- **Tunnel hearing** - I know someone who lost hearing in one ear during their first large process upset. If that was not bad enough, they then thought they might be having a stroke which just increased their level of anxiety.
- **Dissociation** – feeling of detachment from what is happening
- **Coordination may be affected** - A common affect is shaking hands. This can be a concern when you are trying to double-click on a controller and you shake enough that the mouse moves over a pixel and the double click is not recognized. I have witnessed this many times as a Simulator Trainer. Triple clicking can mitigate this (check your control systems functionality to ensure this does not create a new hazard).
- **Speech can be affected** - Someone shared that they lost speech during a process upset and could not communicate with the outside Operator. The laughter of the other panel operator did not help the situation. This can also appear as a high-pitched tone that can be difficult to understand, especially with some of the new digital radios.

Just knowing these are possible should help minimize their impacts if they do happen to you.

"The Unthinkable" does a great job of explaining the three stages that we go through during critical situations. Understanding these can help you recognize when you may be stuck and allow you to move past them. Below is a brief explanation of them.

1. **Denial** - We are looking for patterns from past experiences to associate the event with. There is an effort to normalize the event as we try to fit it into past experiences. This can diminish the severity of the situation.
2. **Deliberation -** People tend to form groups and discuss our situation. We like to socialize and help each other. Nobody wants to be the first to panic and embarrass themselves. We want to see if there is a group consensus first.
3. **The Decisive Moment** - We finally get to the point of taking action, hopefully not too late. Actions to critical events need to be rehearsed and all the implications thought through beforehand. This will move us through denial and deliberation, so we do not have to think through critical situations. Scenario based training is one of the best tools I know of to get to this point faster. This can be as simple as fire drills for emergency response. Scenario based training will be discussed more in the Chapter 5 - Training.

## Strategies to manage stress and anxiety

### Breathing

One common stress reduction technique that is discussed by many disciplines is breathing. Respiration is an autonomic function that we can consciously control and can be used to reduce our physiological responses. Breathing and emotions are directly linked and influence each other. Using breathing exercises can result in a reduced pulse rate.

Tactical breathing is quite commonly used by the military, fire fighters, police, emergency room doctors and now Control Room Operators. The concept of tactical breathing was mentioned in Amanda Ripley's book as well. Like anything else, if you do not train you are less likely to use it when it counts and it will take up more of your cognitive functions when they are needed the most. Two approaches are described on the next page.[15]

**Box Breathing**

**This technique is used to relax yourself by taking 3 to 5 breaths as described below. The technique can be used throughout the day and can help keep you from peaking over the optimal performance curve and having your performance decline when a crises situation arises.**

**Visualize each number as you count and breath in deeply filling your diaphragm and chest with each breathe.**

- **Breathe in counting 1, 2, 3, 4**
- **Stop and hold your breath counting 1, 2, 3, 4**
- **Exhale counting 1, 2, 3, 4**
- **Stop and hold your breath counting 1, 2, 3, 4**

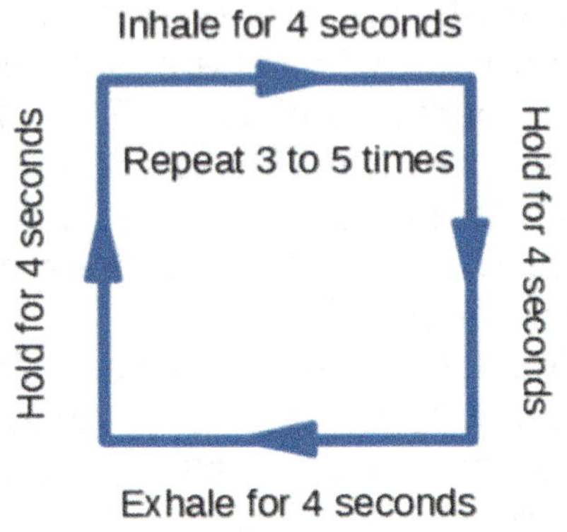


**Don't Crack the Egg**

**This technique is used when you are in a crisis situation. It is a simpler strategy and is based on a 1 to 2 ratio. Inhale through your nose for 4 seconds and exhale for about 8 seconds. The count is not critical. Just breathing will go a long way.**

**This is adapted from Mark Divine from SealFitt. You can also refer to his You Tube video where he steps you through both types of breathing.**

**Don't crack the egg is not a term Mark uses. Hopefully an easy way to remember the strategy.**

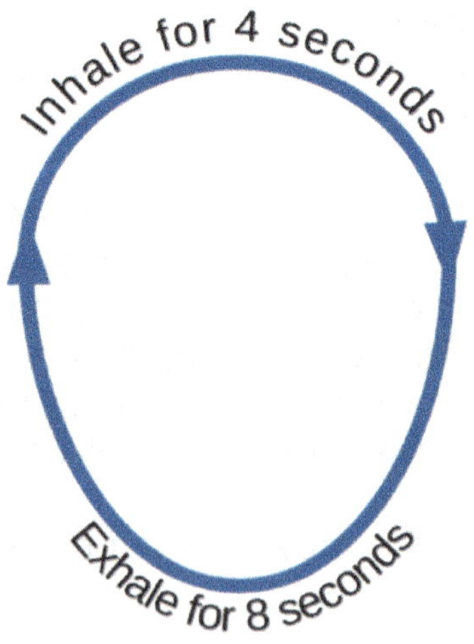


## Anchoring

This technique can help you reduce your stress response and help you to achieve a more relaxed state faster. First, practice the breathing exercises until relaxed. Then perform a movement that you would not normally do to anchor the feeling. When you perform the movement again it will help to induce the same emotional state. The movement could be as simple as touching your small finger and thumb together on your non-dominant hand.

Anchoring can be combined with breathing exercises to compound the benefit.

**Self-Talk**

This is a tool that elite athletes use to improve performance. This could be thought of as the directions you are giving your unconscious mind. Keep the phrase positive. Phases like “I am focused and in control“ vs. “I will not freeze up“. They should be short and in the first-person present tense (I.e., “I am” vs. “He will be”).

Examples:

“I know I can do this; I have trained for a plant trip.”

“My training and experiences will get me through this.”

Be curious about the problem. This will take you back to the present, and you will spend less time wondering how embarrassed you will be if you are not successful.

“This is an interesting problem. I wonder what could be causing it?”

Be patient with yourself if you find yourself experiencing anxiety about your ability to handle situations. If the negative thoughts continue to keep popping up in your mind, let them pass and move onto the visualization exercises. Use these emotions as a motivation to continue practicing these tools and train in your unit.

Befriend the moment and see pressure moments as a challenge or opportunity.

**Visualization**

Use visualization exercises and mental practice. For example, imagine you are in the control room sitting at the console. An emergency alarm comes in, and at first you become quite anxious until you start tactical breathing. Although your senses and attention are heightened you start to calm down. You can clearly see the required information that helps you come to a conclusion about the upset. You decide on the best approach to solve this problem, and you complete a mental simulation which validates your approach. You are surprised by how assured you feel and by the clarity of your thought process. You imagine yourself performing each of the required steps. When you have completed these, you have a sense of relief. You validate that the moves you made are effective. A colleague congratulates you on how well you handled the incident.

This technique can be used for every emergency procedure you may be required to execute. If this is done vividly enough, you will be able to experience some of the physiological effects, and this will provide an opportunity to practice tactical breathing and the anchoring techniques.

The intent is not to be sure you remember every step of the procedure--it is to develop mental networks which will help to minimize the physiological responses. The more realistic you can make the visualization the better. See yourself going through the steps in the setting it would take place (i.e., You are sitting at the control console and the alarms are going off creating a sense of urgency).

Become aware of your physiological responses. You will experience some excitement during an event. The intent of these exercises is to prevent it from impacting your performance. Go through them several times and increase the amount of detail each time.

You could then go through each step of the procedure at the actual console by going to the graphics you would need and selecting the tags you would need to manipulate (without making any actual changes of course). Even better is if you can practice this on a simulator.

### Focus

There may be many demands during a process upset, and it can be easy to be pulled from the main priority. The use of a trigger word can be used to bring full attention to the task or procedure at hand. This will help you to focus on the relevant cues to complete the required task (e.g., Executing an emergency procedure after a plant trip). The word "Focus" could be used or "Execute". This will also help steer away from thoughts of wondering how others are perceiving our performance or dwelling on the consequences if we mess up.

After you have completed the required first steps you may experience a feeling of being overwhelmed, and you may tend to fixate on what just happened and how well you performed. In our industry this is a luxury we cannot afford. There may be secondary issues that need to be addressed. Now it is time to take a step back and regain situational awareness. Using a cue word can be beneficial in helping re-establish situational awareness and to protect against fixation. A couple of examples of cue words could be: "Assess", "Take a step back", "Big picture".

It is commonly felt that we can only focus on a limited amount of information and can only reason through one problem at a time. Dr. Gary Klein's Recognition Primed Decision model helps us to understand how we recognize patterns to quickly assess a situation without having to logically work through every aspect of the situation. This takes familiarity with the situation, either by experience or scenario-based training.

### Hormones

The interaction of hormones is very complex in our bodies. Two hormones that affect how we deal with pressure situations are cortisol and testosterone. To look at them simplistically, cortisol engages the amygdala that allows us to respond quickly to threats with either a flight, fight or freeze response. The amygdala is the part of the brain that is normally associated with a fear response. This is typically at the expense of our reasoning abilities and cognitive functioning.

Testosterone helps to regulate the affects of the amygdala and increases our reasoning abilities and cognitive functioning. Research indicates that more testosterone does not lead to taking "stupid risks". Testosterone helps us to become more rational.

A study showed that adopting a body position that is more open with the shoulders back and the arms open increased testosterone levels by 20 to 25 percent and decreased cortisol by 20 to 25 percent. The effect was found equally in men and women.

Both hormones have their benefits and are quite necessary to allow us to interact with our environment. This information is intended to help us manage our hormones based on the circumstances we are in.[16]

## Deeper Dive - Autonomic Nervous System

Let's take a deeper look into how our body/mind handles life's stressors. Our nervous systems are thought to be divided into the central and the peripheral. The central includes our brain and spinal cord. The peripheral is everything else. My experience is they are intimately connected and respond together. Our Autonomic Nervous System (ANS) controls the hormones that regulate our bodies and how our brain functions.

To keep it simple, we will only look at the main responses of the ANS. The sympathetic system prepares us for activity and threat responses such as fighting or fleeing (fight or flight). Our parasympathetic system prepares us for rest and for the threat response of freezing which may be a more common response. What I have observed and from my personal experience is that we are much more likely to freeze sitting at our control console than to get up and go home. We may act defensively or aggressively if our sympathetic system is overactive.

Normally there is a balance between our sympathetic or parasympathetic systems. If the sympathetic system is over activated, it will place us in a state referred to as hyper-arousal. If the parasympathetic system is over activated, we will be in a state of hypo-arousal.

Understanding what your tendency is may be helpful in recognizing when you are in this state and for you to take measures to mitigate the effects. Focused breathing will help for either state.

**Hyper-arousal:** Fight or flight, reactive, aggressive, limited focus
**Hypo-arousal:** Freeze, disconnected, passive

The window of clarity is about the psychological and physiological setting in which we can regulate after being activated by a stressor. This is the window in which we can effectively make decisions and respond to problems.

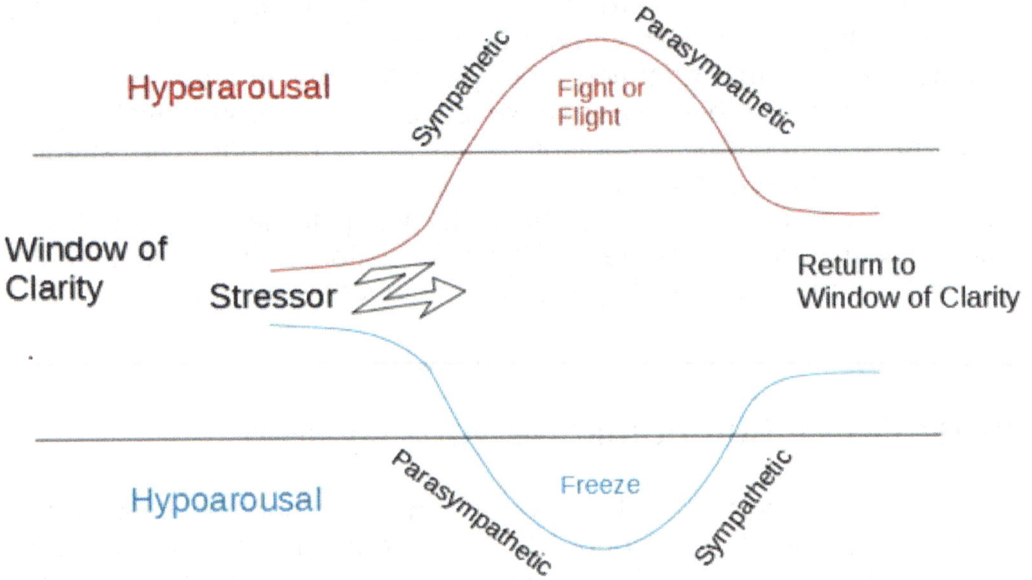


Diagram and notes adapted from "Nurturing Resilience" Kathy L. Kain, PhD, and Stephen J. Terrell, PsyD.

## Self-Regulation

There have been many times, after some event at work, when I continued to work through the problem well into the evening and tried to figure out what I should have done differently. Or there have been times when I was so mentally exhausted that I could not think until I would wake up at 2:00 AM in a hyper-arousal state. This can be part of the learning process, or it can be just keeping yourself in an agitated state longer than necessary.

Whether it is during an event or after, knowing how to self-regulate our nervous system is beneficial. Self-regulating allows us to respond properly to an event and to maintain our mental and physical health. This does not mean we will never exceed our window of clarity. It means we can lessen the amount of time we are out of it. The tools mentioned in this chapter will hopefully assist with this. The other very important piece is to be prepared as much as possible for these events--which is what the rest of this book is about.

Self-regulating also applies to how we react to the day-to-day challenges. If we are continually operating at the edge of the window we will easily move out of our window of clarity when an event takes place.

Responding from instinct can be helpful if we must jump out of the way of a speeding car or if we have trained to respond to a very specific emergency. It is not as helpful when we are interpreting complex situations and making decisions. When we respond from instinct, we are acting from our sympathetic nervous system. Taking deep slow breaths can activate our parasympathetic system and return us to our window of clarity where our reasoning and intuition can function.

In the section on patterns, I mentioned the research lead by Lisa Feldman Barrett and that our emotions are based on the patterns and concepts that we hold. Our emotions are also responding to our prediction of what we believe is about to happen based on cues, and then we update our prediction based on what is really happening. This is something to keep in mind when our emotions are getting the best of us. We may have to ride the wave of the initial adrenaline spike we get, take some deep breaths, and then tell ourselves "I got this".

Based on my experiences, discussions with other Operators and the information I have researched, responding to critical events is as much an internal process as an external one.

## Managing Fatigue

Fatigue impacts our ability to make decisions and limits our motivation. For shift workers, this is typical due to lack of sleep. This means our troubleshooting skills will suffer and the curiosity to see a problem to resolution may be diminished. This can have a compounding effect that can eventually leave you impaired if it is not addressed. It can also leave us moody which may make it challenging to work with others while problem-solving.

Our bodies have natural biological rhythms they follow, which means we prefer to stay awake during the day and sleep during the night. Below are some ways that we can manage this[17]:

**Light.** As soon as you start to sense light your body starts to wake up, even if your eyes are closed. You can wear sleep masks, although I find these uncomfortable after a few hours. I used light-blocking roller blinds; however, I found they leaked light out the sides. To block the remaining light, I simply cut cardboard to fit the inside of the windows, use duct tape on the edges to make it last longer and then wedge them in the window. If possible, darken your route to a bathroom in case you must relieve yourself partway through the day and want to continue sleeping

**Noise.** Let your family know when you are sleeping and to keep noise to a minimum. If it seems appropriate, also let your neighbours know when you

will be sleeping. Using soft earplugs can help, as well as white noise from a fan or air conditioner. You can also buy white noise machines that can generate the sound of wind rushing through trees or a waterfall.
Turn on the do-not-disturb feature on your mobile phone. My phone has a feature that my "favourites" calls will still come through. This gives you the option of getting calls from your family members if there is an emergency (knowing this may help you sleep easier). Put a do-not-disturb sign on the doorbell. This will keep the delivery driver from ringing the doorbell and getting the dog barking.

**Temperature.** Experiment with what sleeping temperature works best for you. Somewhere between 17°C (65°F) and 24°C (80°F) (this seems warm for a Canadian).

It can also help to keep your daytime as free as possible when working nights. The less there is to worry about the easier it is to sleep. Of course, life happens, and you will have to manage this as best as you can. Physical activity and proper nutrition are also worth investigating, as both have an impact on fatigue and performance.

# CHAPTER 5
# Training

*"Tell me and I forget, teach me and I may remember, involve me and I learn"*
*Benjamin Franklin*

When I first started as an Operator, I took a troubleshooting course from K & S Troubleshooting. Two Operators from Houston developed this course and decided to take it on the road. They had some great stories to share which had us all captivated during the course. One point that they kept driving home is that we must know what normal is.

Initially I thought this meant memorizing each process variable and being able to recite the values on demand. This can be handy at times and impress your peers, but if you rotate through all the areas and do not work the panel all the time it can be a challenge. Constantly changing rates can also make this difficult.

I am not that great at memorizing numbers. It takes some effort on my part, and I must recall them occasionally for it all to stick. However, in the patterns that I have developed during my career, I have a general sense of what normal is and when the process is looking odd. When I am viewing the process, my pattern recognition kicks in, and I can detect anomalies. (Review Recognition Primed Decisions in Chapter 1.)

One advantage of modern control systems is that we have trends at our fingertips which can fill in the data that may not be in our memory banks. This is very useful when we are trying to validate when our pattern recognition is telling us something is odd.

One challenge most new Operator's face is that they may have very little to build on to develop patterns. They may feel like they have just moved to another planet. Developing a sense for what is normal can take some time, and I have seen some new Operators become overwhelmed by the complexity and the size of everything.

It is one thing to become familiar with the 5 HP compressor at your parent's garage, but it is a large step up to now be expected to operate a compressor as large as a garage with a 50,000 HP driver. I had the benefit of working as a labourer during the construction of the first facility I worked in and saw it built from the ground up. Not that I understood what everything did, but it was familiar when I started to operate there.

At the second facility that I operated at, I was there for commissioning and start-up. I helped to write procedures and develop the control system graphics. I was also involved in the testing of the simulator. That gave me a head start on knowing what normal was. As a shift technical leader (Shift Lead), I wanted to know everything I could learn before we started up the facility.

Understanding the extent of your responsibilities is a good motivator. Knowing that one wrong move could take the process down and result in a million or more dollars in losses is one thing. To fully comprehend that a decision you make or do not make could harm someone or result in the loss of life is another.

Typically, these disasters are not the result of one problem or error. This likely will be of little comfort if you lost someone on your crew. I am not intending to overwhelm you with the extent of your responsibility. I am trying to highlight the importance of training and understanding the impact your decisions will have on the process and the people you work with.

You need to be confident that you understand the principles of how the process works and that you don't stop looking until you do. Ask yourself, am I going to feel comfortable operating this equipment if I do not understand it?

For an outside Operator to understand normal they must look, listen and feel. A healthy curiosity is the first step to building patterns that will help you detect when something is unusual. There are always noises in the unit. You should be able to notice subtle changes that may indicate a new vibration, process, or mechanical problem. Another example is when a pipe is frosted that does not normally have frost on it.

A Panel Operator needs to spend time reflecting on what each controller does and how that impacts the process. Knowing the limitations of the system's pressure, temperature, and product specifications is a must. I highly recommend a reading sheet that can be filled out by hand as an excellent way to get to know normal (there is an example in the Rounds and Checks section, Chapter 7 of this book). These were used extensively in the past, and this gave Operators a controller's output at their fingertips in case the transmitter failed. They then could quickly return the controller to the previous output to minimize the upset. They have become less common because trends can quickly call up history. This can lead to the Panel Operator becoming less intimate with their panel.

In addition to this, a Panel Operator needs to spend time reviewing the graphics to keep current on the process. This would involve questioning what each controller does and reviewing the whole graphic to ensure they still understand what the obscure number in the far corner is telling them. This will lighten the cognitive load during an upset.

Most people tell me they learn better by doing rather than by reading, and they are right! We are so much better at building patterns through in-depth experiences than by memorizing information from manuals and procedures. This does not mean that some basic understanding of how the process works is not required before you can begin to develop patterns of how the system works.

A term that can be used to describe pattern development is **associative learning**. This is where a person mentally links concepts and ideas. These links then can be applied to other linked networks to allow us to apply that knowledge to similar situations.

During a process upset, we do not need to have memorized every possible situation or to think logically through every possibility. We can make inferences from similar experiences to a new situation based on the associations.

Humans can build on associative memories. Once we have a foundation of the basis of our systems, we can apply that knowledge to other systems. For example, once you understand how a PID controller works you do not have to memorize the same information for every controller (See Chapter 7).

Through associations, you can use that same information of every system that uses a PID controller. From there you can bridge to associations of how a PID controller is used for level control or flow control, etc. These associations then can be broadened to how the controllers affect the equipment they are connected to. Further associations can be made between different systems and how they affect each other. If you fail to make these associations, you will not be able to recognize the changing patterns when the process is upset.

Decision-making is a combination of pattern recognition and analytical thinking. Both skills must be developed. In our first chapter, we discussed how our subconscious mind can sift through large amounts of information and recognize patterns and the conscious mind provides direction and seeks clarification.

To build the pattern recognition we must first go through the step-by-step review of our process and equipment function. Developing pattern recognition does take some work. An experienced Operator will use less brainpower than a new Operator. A new Operator must run through several possibilities to come up with a decision. An experienced Operator's neural associations and pattern development are much more efficient.

## Pattern Development

The concept of chunking has typically been used to make it easier to memorize information. A common technique is to group items that are similar or relate them to something we already know. For instance, if I wanted to remember that someone was born in 1957, I would think of them sitting in a 1957 Chevy. This is effective for me because I like the 1957

Chevrolet. Making it visual helps to tie it into already established patterns. We are taking this a step further and will chunk patterns.

The first step is to understand the basics. It would take quite a large book to cover all the basics of every process that there is to date. I will use a few as examples. There are several types of control valves and positioners. The intent here is not to cover them all but to show a progression of knowledge and developing pattern chunks.

You could argue that you do not need to know all the details of how something works to be able to operate it. If I open a valve, there is more flow; if I close it, there is less. That information may be enough for normal operation. It will not be enough when the flow starts to oscillate erratically, and you do not know if it is a valve problem, a transmitter problem or the many other things that can cause this.

The diagram below shows most of the major parts of a basic control valve with an electro-pneumatic positioner.

The positioner receives a 4 to 20mA control signal from the controller and an i/p converter adjusts the positioner to that signal. The positioner adjusts the air pressure to the diaphragm to move the valve. There is an internal spring that is working towards the closed position, and the diaphragm works against the spring to position the valve. There is a feedback arm in the positioner to allow adjustment of the air to the diaphragm to ensure the valve is in the correct position.

Some valves will fail open or lock on a loss of signal. This valve fails closed on a loss of signal.

Now knowing the basic function of this style of valve we can recognize some failures:

- A loss of air supply to the valve will allow the spring to close the valve
- A loss of the electric control signal to the valve will allow the spring to close the valve
- Pneumatic positioners have many moving parts. If they become worn the position may become erratic, causing the valve to be offset from the requested output. They could also fail the valve closed or open.
- (There are more possible failures than described here.)

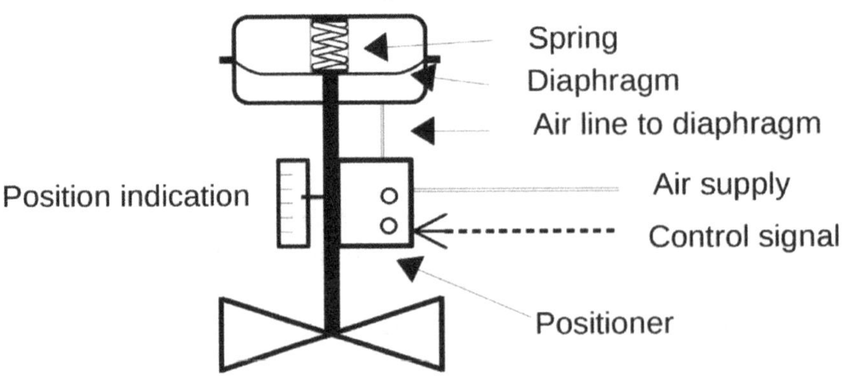

The next step will be to set up the first layer of pattern chunking. This is required so your unconscious mind is conditioned to think through a problem response and the conscious mind does not have to be taxed doing this when many things are going on. In the table below there are two examples. These can be expanded to include more failure possibilities. Microsoft Excel can be used to automatically answer these questions by inserting Option buttons in Active X controls. Most learning management systems can be used as well. At this stage, this may seem very simplistic. It is important to think through the failures rather than just reading them.

| Failure mode | Valve Opens | Valve Closes |
|---|---|---|
| Wire to positioner is broken | | X |
| Instrument air supply to valve is lost | | X |

This can be the first stage of your understanding. The next step would be to understand how the tuning on a controller affects the output. (Refer to the Control Basics section in Chapter 7 for more details).

## Control Systems

Now let's look at the valve in a basic control system.
The diagram below shows a level to flow cascade loop:

- The level transmitter uses the differential pressure from the bottom of the tower (H) and the top of the tower (L). The lower sensing line has higher pressure because of the liquid in the tower. Both sensing lines are impacted equally by changes in tower pressure.
- The flow transmitters use an in-line orifice to create a differential pressure with the higher pressure being on the upstream side of the orifice. The pressure differential is then used to calculate a flow measurement.

Some possible failures:

- The tubing to the level transmitter can plug or become restricted causing the indication to read high or low.
- When the tower level is above the low-pressure tap on the level transmitter it may fill with liquid. This will cause the pressure on the low-pressure side of the transmitter to increase and indicate a lower level than there is (this would not be the case if capillary tubing is used).
- Fouling or debris on the upstream side of the flow orifice will increase the differential pressure on the flow transmitter causing the flow to read higher than it is.

- The tubing on either side of the flow transmitter can become restricted or plugged giving a falsely high or low reading.
- The electronics of any transmitter can fail to send no indication (bad PV).
- A restriction in any sensing line can cause the pressure to the transmitter to have a delay. This may look like a controller tuning problem.
- Large rate changes may result in the valve output operating near being fully opened or closed. This can change its flow characteristics resulting in the tuning no longer being effective.

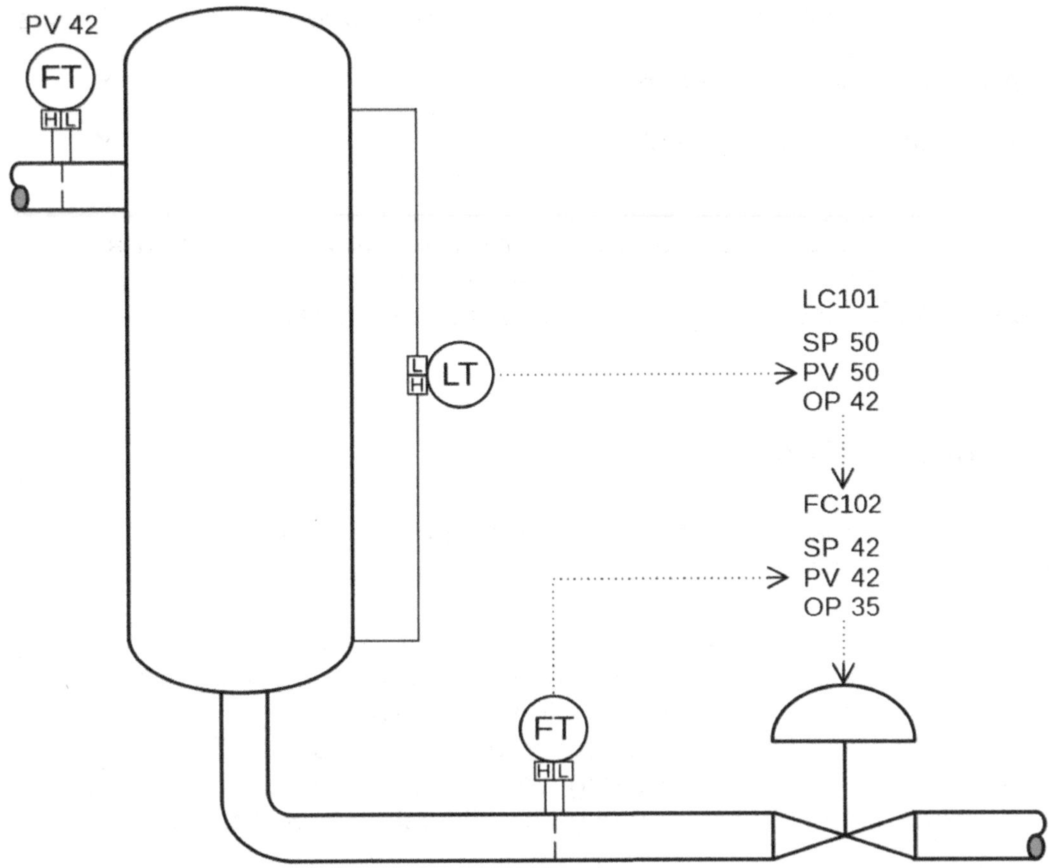

We will now go to the next level of pattern chunking exercises starting with the flow transmitter and then working into the level to flow cascade loop.

| Condition | FC102 output increases | FC102 output decreases |
|---|---|---|
| FT102 sensing line is fouled and reading low. | X | |
| A piece of debris is stuck in the FT102 orifice | | X |
| A sensing line is plugged on LT101 causing a high indication | X | |
| The tower level is higher than the LT101 low-pressure leg | | X |

In the next diagram, we have expanded the amount of equipment to include everything in a distillation system.
This consists of:

- A flow controller (FC2101), that feeds the tower.
- A temperature controller (TC2115) that is located on one of the trays, cascades to a flow controller (FC1051). This regulates the amount of steam supplied to the reboiler.
- A pressure controller (PC2113) removes the overhead product from the tower as a vapor.
- The overhead condenser has a level controller (LC2120) on the refrigerant.
- A reflux drum level controller (LC2130) that cascades to a flow controller (FC2141).
- A level controller cascaded (LC2110) to a flow controller (FC2121) to maintain the level at the bottom of the tower like the last example.
- There is also a pressure differential indicator (PDI2117) that would indicate the differential across all the trays.

A basic understanding of distillation theory would be required as well. There are several books on this topic and information online if your organization has not provided training on the topic. It is not my intent to duplicate all this technical information in this book. The intent is to demonstrate a learning system, not to teach process information. The values in these examples are not based on an actual process. They are for demonstration purposes only.

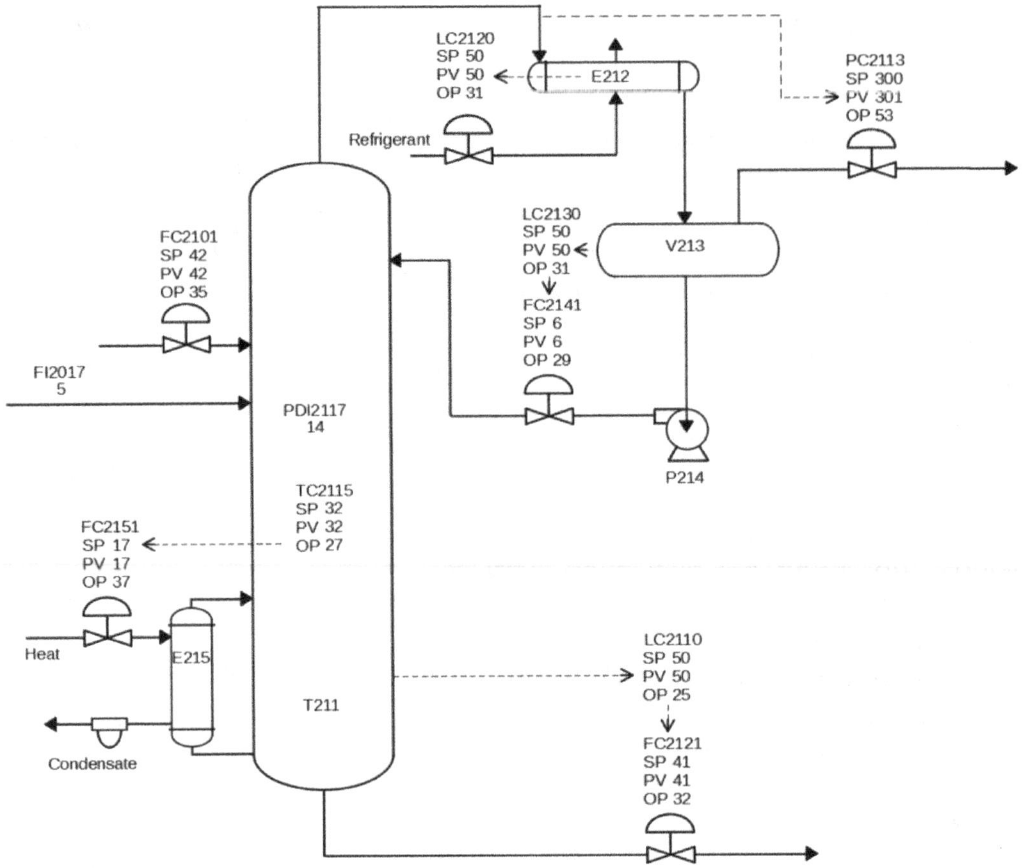


Now that we have expanded the amount of equipment, we can increase the complexity of the exercises that develop the patterns our unconscious mind can use to problem solve during an upset. Even though this can seem repetitive and boring, it is necessary to develop the neural connections that patterns require.

In the following table the impact to the controller's outputs would be filled in with either, Increase, Decrease or Same. We are looking for the initial response of the controllers. They may eventually adjust to the new condition.

| Condition | FC2151 | FC2141 | LC2120 | PC2113 | PDI2117 |
|---|---|---|---|---|---|
| 1. An increase in flow from FC2101 | Increase | Increase | Increase | Increase | Increase |
| 2. FC2101 flow is indicating high | Decrease | Decrease | Decrease | Decrease | Decrease |
| 3. FC2151 transmitter indication is failing low | Increase | Increase | Increase | Increase | Increase |
| 4. PC2113 fails open | Decrease | Decrease | Decrease | Decrease | Increase |

A short description can also be given after, to help understand the responses if the trainee is struggling. How this is done will depend on whether you are using software or paper.

For example:

Referring to Condition 3 above, the controller will increase the output and more steam will flow to the reboiler. This will increase the vapor traffic up the tower driving some of the heavier components up the tower, which means more inventory in the upper section and less in the bottoms. This will likely cause the overhead product to go off-specification as well. TC2115 will eventually start to reduce the set-point of FC2151 in response.

The next step will be to look at systems that are directly tied together and can impact each other. Understanding how systems that share a pressure envelope respond to pressure changes is one example. The refrigeration system mentioned previously with several users is another good example. If the suction pressure changes on the compressor this will change the temperature of the refrigerant and impact the performance of the systems using the refrigerant.

In this example, I have added a pre-stripper to the previous example. T201 does a rough separation and since it does not have a reflux system the overhead vapours are further separated in T211. In a system like this, an upset in one tower will impact the other.

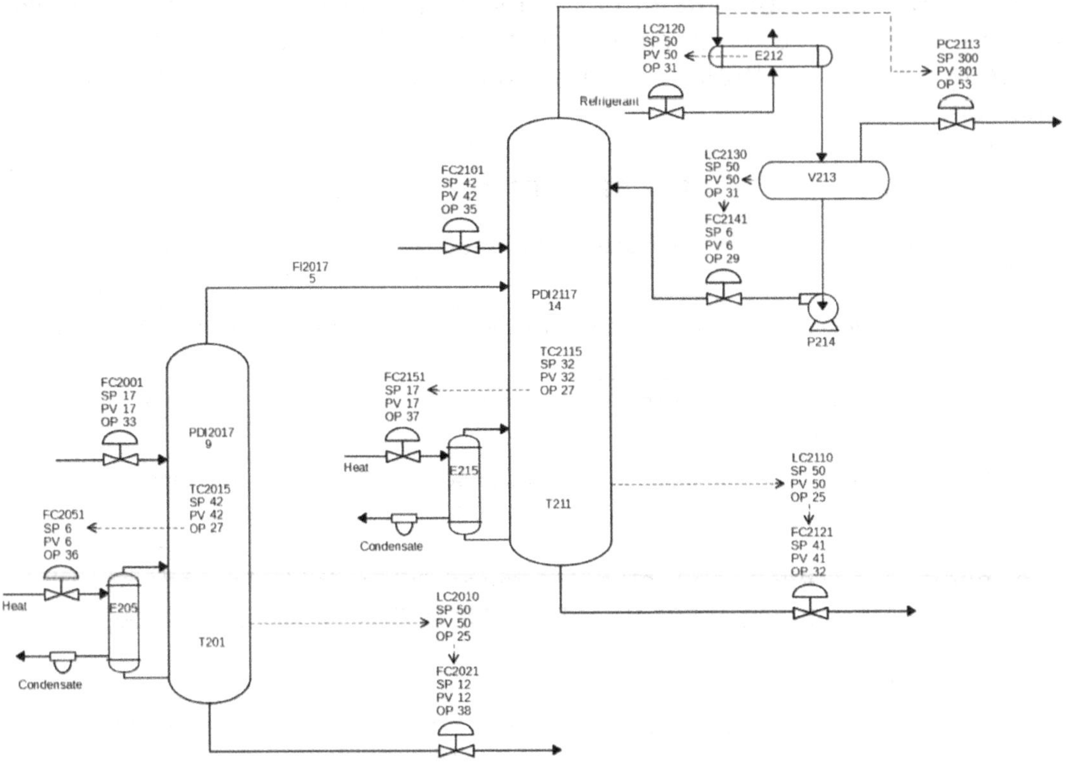


## Example questions:

| Condition | FI2017 | FC2021 | FC2141 | PC2113 | PDI2117 |
|---|---|---|---|---|---|
| 1. A temperature increase on TC2015 | Increase | Decrease | Increase | Increase | Increase |
| 2. PC2113 transmitter drifts higher | Increase | Decrease | Increase | Increase | Increase |

The next step would be to create a diagram that includes most of the process and identifies the major control points. This would "chunk" the process into manageable pieces. It can be a simplified diagram that does not need every transmitter or controller. The intent would be to include the major flow controllers, pressure controllers, or level controllers that impact more than one system. FC2012, FC2121, and PC2113 in the previous diagram are examples of these. This will help to develop a mental model of the process and then develop it into a dynamic mental network. A similar questioning process for reactions and upsets as those shown previously should be developed as well.

# Scenario-Based Training

**The difference between scenario-based training and traditional training.**

Traditional training is designed to help with the memorization of facts and procedures. This is an important first step. Scenario-based training is a form of experiential learning. It helps the learners to look at the process in new ways and to see the interconnectedness of the different parts of the process. It helps the learner develop more sophisticated mental models, build tacit knowledge, and develop pattern recognition. All of these are keys to making decisions when there is limited time. Regardless of how you make decisions, scenario-based training will help.

Scenario-based training is designed to mimic real events or likely events. It develops "common sense" and prevents what has been termed as "book smart". Rather than asking the trainee how to respond to a known problem, a set of circumstances is given, and the trainee must first identify the problem and then provide the response. The pattern chunking exercises build a basis for solving problems. We have developed the skills to recognize cause-and-effect within the system. The next step is to learn to recognize problems and decide how to correct them.

On the next page, we have two malfunctions. For both malfunctions, the first alarm would be the PV of FC102 reading low. In both cases, the output of the flow controller has ramped open to 100%. This output value is the signal to the valve, not feedback from the valve. The difference is the level in the tank is rising in the first example and dropping in the second.

For scenario-based training, the trainee would be given one of these examples and asked to identify what the problem is and then how to correct it.

In the first example, the control valve has failed closed and would require the bypass to be opened if there is one. In the second example, the flow transmitter has failed low, and the valve would have to be placed in manual and the output returned to where it was before the failure.

Also, notice the flow into the tank has not changed. This piece of information can be used to validate that the problem is on the outlet of the tank. Although not included here indicators in the equipment downstream of this vessel could be used as well.

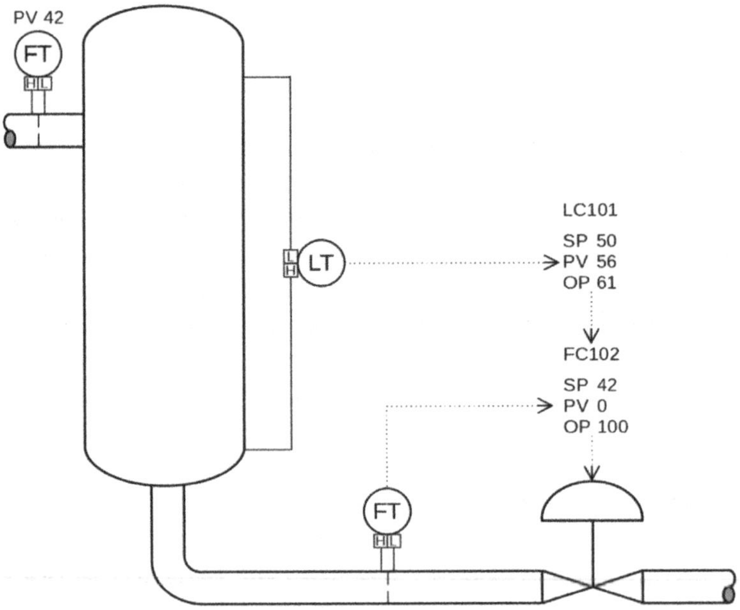


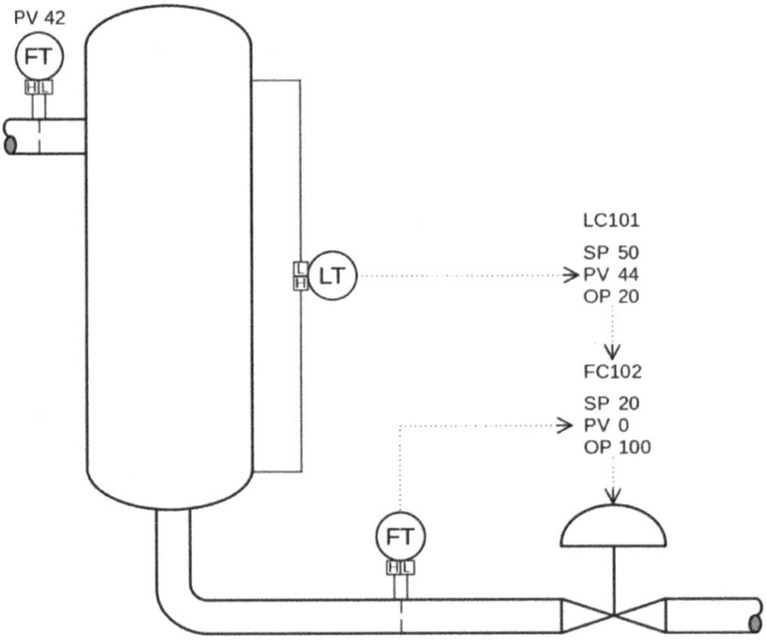


**Benefits of scenario-based training:**

- Increased retention of information - Applying knowledge in a situation helps promote pattern development and integration with existing patterns.
- Development of tacit knowledge - This is the knowledge based on experience and allows us to respond to situations when there is time pressure.

- More engagement - Interactive training draws in the trainee versus passive training where the trainee is only listening or reviewing information.
- The development of the scenarios will draw out the experience of the experts in the field.

The following are some examples of scenario-based training:

**1. DMX (Decision Making Exercises)**

These were developed for the COP (Center for Operator Performance)[18] by Macro Cognition (Dr. Gary A Klein). This training was adapted from the Tactical Decision Games that John Schmidt had developed for the US marines.[19] The trainee is given a set of process conditions and first has to detect the problem and then explain how to fix it.

A manual was created on how to develop this style of training which includes the cognitive elements of decision-making. DMX is licensed by Macro Cognition and is available for use by COP members.

**2. ShadowBox**

A benefit of this training is it breaks down the common decision points used to work through problems. For trainees, this can be helpful as they may not know how to begin to solve a problem. ShadowBox training is comprised of three main components: scenario, decision points, and expert feedback. The training presents challenging scenarios through text, images, and/or video. At various points throughout the exercise, the scenario is paused, and the trainee is required to answer a decision. These decision points often ask the participant to reflect on the situation up to that point. This might include prioritizing a list of options (e.g., what is most important to you), or diagnosing what they think is wrong, or anticipating what could happen next. Each decision point taps into the cognitive dimension of the problem, and makes the trainee reflect on their knowledge, goals, and expectations. As part of the decision point, the trainees must write down the rationale for their decisions. Finally, after each decision point the trainee gets to see how a group of experts responded to the same prompt. At this point, they get to compare their own responses with the experts' responses and capture insights they might have missed.

ShadowBox training bypasses the expertise bottleneck that many organizations experience. By capturing the expert feedback upfront and packaging it within the scenarios, the experts are not required to sit in and/or facilitate the training. Rather, trainees can complete the scenarios independently or as teams on their own time. Alternatively, trainers can lead ShadowBox training exercises with individuals or groups to enhance engagement and learning.

ShadowBox is also very good at capturing the knowledge of experts. Experts take part in the scenario creation process. As previously mentioned, an expert may not know how they know something or how they arrived at a decision. Developing these scenarios allows them to reflect on how they have made past decisions and break them down into decision points.

There is a Web App available from ShadowBox LLC. A training manual was developed for the COP and is available to COP members or through ShadowBox LLC at www.shadowboxtraining.com.

The original concept was developed by Neil Hintze who was a Battalion Chief with the New York City fire department. He developed the training to capture the knowledge of expert firefighters after the 911 disaster in New York. Dr. Gary Klein and his associates have continued to develop this style of scenario-based training and it is now licensed through ShadowBox LLC.

**3. IADs (Immediate Action Drills)**

This type of training is used when there is a known sequence of emergency steps required in a short time period. Because of the consequences and time pressure, our cognitive functioning may decline during these events. Developing these automatically fixed action patterns offload our cognitive functions and allow us more capacity to monitor the situation. These could be the steps required to isolate the process after a power failure, steps to stop a reaction, or even the response to a release to the atmosphere. Adding time limitations to these drills will help you prepare for an actual event.

**4. Simulation training**

This is likely the best way to practice responses to process upsets and to develop troubleshooting skills. A dynamic, high-fidelity simulator that uses the same interface as the actual control panel can provide very realistic training.

The simulator is only a tool and developing all the previously mentioned skills will make it a successful tool. Like any tool, you can use it to bang someone on the head or help them develop the skills required to operate our very complex processes.

I have qualified on two panels without simulator training and two with simulator training. I look back on the skill level I had on the first two panels and wonder what my performance would have been like if I had to respond to a major event. I was proficient in the day-to-day activities but was never really tested because of the reliability of our process. Being able to practice upset responses and problem-solving with a simulator on the next two panels helped to build my competency faster. It also helped to build my confidence which led to better decision-making when I had to respond to an upset.

There is a reluctance to purchase simulators and develop a program because of the cost. I believe the cost is far outweighed by the benefits (see Simulators in Chapter 7).

# Learning From Mistakes

"If you can't be smart in hindsight, you can never be smart."
Dr. Gary Klein

Being an effective trainer takes humility. Realizing that when you are trying to explain something, and they are not getting it, it may be just as much your shortcoming as theirs. One of the characteristics of tacit knowledge is we have difficulty explaining why we know something. As experienced Operators, we have lots of tacit knowledge and it can be very difficult for us to understand why the trainee is not getting something we believe is obvious. It is our job to figure out how we know what we do. What are the cues and patterns that we notice that explain how we understand the situation?

Judging the performance of a trainee too early can lead to excess stress and poor performance, as discussed in the Chapter 9. Distinguishing between a learning phase and a testing phase will help both the trainee and trainer. There is value in letting trainees make mistakes. The knowledge tends to stick a lot better. This concept applies to discussions and simulator training, not in an operating unit where the consequences of a mistake are not acceptable.

I believe we have just started to develop effective training. We need to be constantly exploring new ways to help the trainees understand concepts and make decisions.

In his book "Streetlights and Shadows, Searching for the Keys to Adaptive Decision Making"[20] Dr. Gary Klein discusses ten surprises about how we handle ambiguous situations. One of the concepts that he challenges is the idea that all that is required to help someone improve is to give them feedback on what they did wrong. This may be enough for simple tasks. It is insufficient in our complex processes.

We have to try to understand the problem from the perspective of the trainee. What are they seeing? How are they perceiving the information? What are we assuming they should understand but do not? What do we know from our experience and find difficult to explain?

The diagram below is adapted from Sidney Dekker's book "The Field Guide to Understanding Human Error".[21] This is another book that relates directly to our industry and is well worth reading. If you are the person on the outside of the tunnel you have the perspective of seeing the entire sequence of events and the triggering conditions.

If you are the person directly involved, you are the person inside the tunnel. The outcome is not known, or you may have made a different choice.

An Operator shared with me after a moderate process upset that they could have noticed the problem sooner and reduced the impact if they

would have been on the right graphic. They had just left this graphic because someone behind them asked them to change graphics.
Think of all the demands an operator may have, (alarms, phone calls, radio chatter, conversations in the control room) that can divert their focus away from critical information. Each change in the direction of the tunnel is caused by the decisions of the person going through the events based on external stimuli.

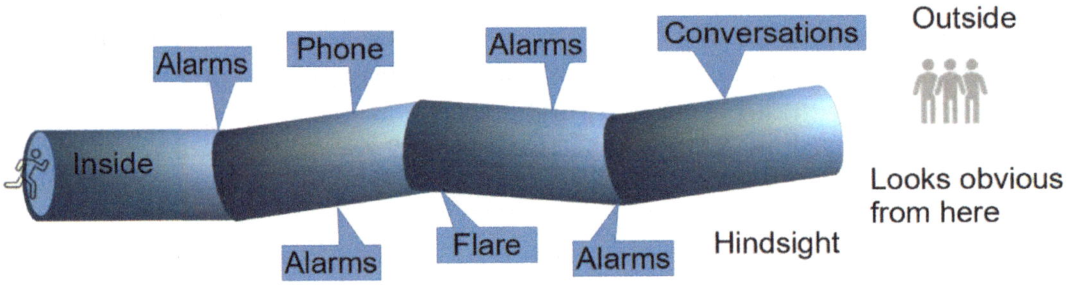


Diagram adapted from "The Field Guide to Understanding Human Error" Sydney Dekker

## Training Others

"Knowledge is great, procedures are helpful, decision-making skills are essential"
Author

Traditionally training means trying to get someone to memorize information, teach them rules, and follow procedures. This strategy makes sense for simple tasks that can be broken down into steps. A training formula can be set up that can be easily followed.

We have already covered why this approach is not enough. Not only do we have to develop our cognitive skills, but we also must create an environment that will foster these skills. The development of cognitive skills will lead to better decision-making, accurate sensemaking, more creative improvising, and faster detection of problems.

## Cognitive Coaching

Dr. Gary Klein wrote an essay for Psychology Today called "Cognitive Coaching' Six mindset shifts trainers should make"[22]. This was built around discussions with Joseph Borders during our mental model project with the

Panel Operators at our petrochemical facility. This project was completed for the Center of Operator Performance.[23]

The first section examines six mindset shifts that trainers should adopt. The second section provides suggestions for trainers about what to watch out for as they adopt the new mindsets.

The following is taken directly from Dr. Gary Klein's essay. It was written in clear and easy to understand terminology and I did not see the need to alter it.

---

***Start of essay***

**From criticism to curiosity -** The belief in question is that the job of the trainer is to spot mistakes and correct them right away. One senior trainer told me that this was his mindset starting out, but over the years he had learned to be curious. Early on in his career as a trainer, if someone made a mistake, he would happily rebuke them, which felt good but didn't seem to be achieving very much; over time his mindset shifted and now if someone makes a mistake, he wonders why and tries to get to the bottom of it, working with the trainee to figure it out. This new mindset is making him much more successful. The people he is training make much faster progress. Too many trainers are just annoyed by mistakes rather than seeing them as opportunities. These trainers interpret mistakes as signs that the trainee wasn't paying attention, or didn't care, or maybe that the trainer did a poor job. In any event, with the criticism-oriented mindset, mistakes are blemishes. However, the curiosity-oriented mindset views mistakes as opportunities. If the trainer adopts a curiosity-oriented mindset, he/she can hopefully figure out what is confusing the trainee and help to clarify the trainee's thinking. Trainers should be seeking to harvest mistakes, not avoid them or make trainees feel guilty about making them.

**From following procedures to gaining tacit knowledge -** The belief in question is that virtually all jobs can be boiled down to procedures and steps and the job of the trainer is to teach these steps. While procedures are important and necessary, they aren't sufficient. Expertise is not a matter of following steps. Just about all challenging tasks have too many contingencies and possibilities to be handled by procedures. Expertise is about what we do when the procedures don't clearly apply — when we must use our judgment. Expertise is about gaining tacit knowledge — knowledge that can't be captured by procedures and steps. Tacit knowledge includes noticing subtle cues, seeing patterns, spotting anomalies, and using our mental models of how things work and how they don't work. Trainees want to buy into the procedure's belief — they want to take refuge in learning more and more procedures, adding them to their toolbox, trusting that procedures will take care of just about any problem they might face. The trainer must help the trainee get beyond this procedural mindset and move onto the path to expertise.

**From getting through the material to encouraging curiosity -** The belief in question is that most of what the trainee needs to learn is in the PowerPoints and lecture material. Therefore, the trainer must make sure he/she gets through all this material. Class discussions and questions just chew up time and get in the way of covering the material. As a result, trainers often unwittingly block curiosity on the part of the trainees. They squash curiosity in a few ways (I am indebted to Ed Noble at Nova Chemicals for these observations): They discourage questions. They use ridicule, even with facial expressions, to make trainees afraid of looking dumb. They overwhelm the trainees with details. They introduce so much complexity so early that the trainee stops being curious and retreats into a passive stance of desperately trying to remember most of the material being shotgunned. While there are no easy ways to foster trainees' curiosity, you can see that there are many ways to stifle curiosity. Good trainers have learned to control their tendencies that diminish the curiosity of their students.

**From providing thorough explanations to providing focused explanations -** The belief in question is that the trainer should be lecturing trainees with all the details and contingencies that might come up. As we saw in the previous paragraph, excessive details and complexity can stifle curiosity. Good trainers have mastered the skill of diagnosing the reasons why trainees are confused or surprised. Instead of spewing a canned lecture the good trainers can diagnose why the trainee is surprised — what belief has been violated, what assumption has been called into question. Then the good trainers can speak directly to that belief, that part of the trainee's mental model that needs adjustment. The trainers are asking themselves, "What assumption are these trainees making that is hanging them up?" Good trainers can provide a focused explanation instead of a comprehensive one.

**From explaining to discovering -** The belief in question is that the job of the trainer is to explain things. Often that is the case, but trainees learn more when they discover things for themselves. The challenge for trainers and coaches is to craft exercises and experiences that enable discoveries. That is much harder than just offering an explanation.

**From evaluating to training -**The belief in question is that during the training period the trainers should assess which trainees are struggling and should be dropped from the program — the trainers can save everyone a lot of effort by culling those who are not a good fit. The problem with this approach is that it severely interferes with training, as I explained in an earlier essay. If I know you are watching me to see if I measure up, I am going to be super defensive and guarded during training. I am going to be careful not to get caught doing anything stupid. I am NOT going to be exploring different strategies. I am NOT going to be eager to get feedback on what I did wrong. If I do get critical feedback, I am going to try to make excuses rather than take it in. By combining training with personnel evaluation, organizations interfere with the training process. Organizations do need to evaluate performance. Rather than mix evaluation and training they would do better to set up special test trials and tell the trainees which these are. In this way, trainees know when they have to be on their guard and when they can relax and learn as much as possible.

## Section Two: Seeing the invisible

If expertise is based on tacit knowledge and tacit knowledge is, by definition, hard to describe or notice, what can trainers and coaches do? Here are a few suggestions.

**Subtle cues-** Be on the lookout for visual and auditory and even tactile cues (e.g., mechanical vibrations) that the trainees need to notice. It's hard for trainers, for experts, for anyone, to clearly explain what subtle cues they are picking up on. That's why trainers find it so tempting to simply explain the procedures. But these subtle cues are critical, so trainers can try to be alert to them during operations and bring them to the trainees' attention. Even if the trainee can't make a perceptual discrimination, knowing that experts can make this discrimination can motivate the trainee to practice. (See the work of Anders Ericsson on deliberate practice). One Navy Electronic Warfare Coordinator I interviewed explained that in his early career he couldn't distinguish the signal profiles of two large aircraft — it might have been a KC-135 and a Boeing 737 (I don't remember). He knew that the experienced EWCs could tell the difference in signal signature, and so he became determined to do the same. Therefore, whenever an airplane of that approximate size appeared he became extra vigilant trying to predict which it was, a KC-135 or a Boeing 737, and then he used other sources to check on his judgment. In less time than he expected, he acquired the skill he needed.

**Hindsight perspective-** Another exercise for identifying relevant cues is to use an After-Action Review of a situation, regardless of whether the trainee handled it well or poorly. The trainer can ask, "With hindsight, what should you have been paying more attention to?"

**Anticipating-** Instead of hindsight, what about using foresight. Many trainees get captured by what is happening right in front of them and fail to think ahead. Trainers can pause the action and ask "What is likely to be going on in x number of minutes?" In other words, the trainer can try to encourage a prospective mindset, a habit of mind to imagine what might happen next, either a likely turn of events or a dangerous turn of events.

**Shifting focus-** Here is another mindset shift the trainer can try to foster — seeing the big picture. Trainees often get captivated by the micro-details of what they are working on, tunneling into the specifics, and forget to look up and take stock of what else is going on. We see this tendency in aviation, in operating a petrochemical plant, and in other domains. The skilled decision makers have learned from experience, often bitter experience, to shake loose of the micro-details and come up for air every now and again. This shifting focus is another habit of mind to try to inculcate. The trainer might be on the lookout for times when the trainee is staying micro for too long and either encourage a temporary shift in focus or, if possible, introduce an event in the scenario or simulation to "punish" the trainee who failed to scan the big picture frequently enough.

**Fixation-** We all have the tendency to fixate. When we are trying to make sense of a confusing situation, we usually cast about for explanations and the first one that seems plausible captures us. Instead of telling students to keep an open mind, which is actually bad advice, trainers can be on the lookout for times when a trainee jumps to a conclusion and then holds on to it even though it is incorrect. Remember — there's no shame in jumping to the wrong initial hypothesis. Fixation is a problem when we hold on to that incorrect hypothesis despite mounting evidence that it is wrong. Fixation happens when we explain away the anomalous evidence rather than becoming worried or at least curious. So, the trainer can try to cultivate a habit of mind to become curious about anomalies instead of discarding them. If a trainee is going down a garden path, impervious to warning signs, the trainer can note all the warning signs the trainee missed and use the After-Action Review to let trainees discover how they blinded themselves. Trainees who have just messed up are usually very open to tracing where they went wrong.

***End of essay***

---

The COP (Center of Operator Performance. ) also conducted a training survey with the help of Dr. Beth Blickensderfer from the Embry-Riddle Aeronautical University. One of the findings was that new Operators did not feel comfortable training others. On many sites, a new operator may be training someone a few months after they have qualified in an operating area. This should not have been a surprise as they are seldom provided with training on how to train someone.

The COP has developed some training tools in addition to the scenario-based training previously mentioned. CAARGO (Cognitive After Action Review Guide for Observers) was developed to enhance the skills of trainers that complement any form of training. There is also a companion video to demonstrate these techniques.

Another tool developed by the COP was a workshop on fixation to identify when it is occurring. It also presented tactics and suggestions for the prevention and correction of diagnostic errors. Both these tools are available to COP members and were developed by Dr. Gary Klein and his associates at ShadowBox LLC. They can also be contacted about the use of this guide.

## Simulator Training

As a simulator is a very powerful tool, it would be worth covering some best practices for using it as a training tool. Everything that was mentioned above would apply to simulator training as well. Like any tool, you can use it to help someone, or you can use it to hit them over the head.

In David Strobhar's book "Human Factors in Process Plant Operation"[24] he talks about the three key tasks required of a Process Operator. These are detection, identification and compensation.

It can be easy to focus on the compensation aspect during simulator training. This is when we have the trainee follow the steps in a procedure. This is an important skill.

The detection and identification skills are as important to know as when to implement the compensation step and to be able to handle situations where the cause is not known.

Early detection will prevent the problem from escalating and identifying the correct source of the problem will lead to the correct actions or compensation. The training tools mentioned in the previous section are designed specifically for this.

**Training steps**

Below is the sequence that I follow for simulator training. I have provided some rationale for this order.

Before Operators begin simulator training, it is best if they are familiar with the control systems and can navigate the graphics. This can depend on the availability of the simulator and the trainer. It will also depend on how much time the trainee can dedicate to simulator training.

Beginning with starting up all the systems is a great way for the trainee to work their way through all the controllers, and it allows them to build some confidence. Hitting them with malfunctions right from the start can be overwhelming.

Having them review the start-up procedures with a qualified Operator before the training is helpful and then again just before going through them on the simulator. Highlighting the key steps and explaining the intent of each step can make this easier to manage.

Understanding the reasons for each step and being able to think it through is preferred to only following each step of the procedure. This will prepare them to think through problems as there typically is some sort of failure during a start-up.

Running through the emergency procedures next is a good progression. These are events like power failures, instrument air failures, or large compressor trips. Anything that has a large impact on the process and the results are quite predictable.

Again, understanding the reasons for each step and chunking up key steps can be helpful. There are likely similar steps in several of the procedures. Thought should be given to what should be memorized and which steps in a procedure can be used as a reference. Generally, the first five to ten minutes in a major process upset require immediate response and should be memorized. This will depend on your process.

The next step is to go through situations that are not proceduralized that will require some thinking and problem-solving. Start with basic malfunctions and evolve to the more complex. Even beginning with how to determine the difference between a valve failing closed compared to the failure of a flow transmitter is beneficial. They may initially look the same from the graphics unless you have feedback from the valve to the control room.

Ask them to verify their initial assumptions. Are there indications upstream or downstream of the problem that can validate theories? As the simulator trainer, you will also be acting as the Field Operator and will have to respond to their requests for a field check. I typically deliver the answers from the perspective of a recently qualified Operator and require that I must be asked to check everything and do not volunteer any information.

The expectation is that these are growing experiences and they may not be successful on all of them initially. It has already been mentioned that it is important to distinguish between when they are training versus when they are being tested. Be clear on what the expectations are.

They should be able to start the process up and manage all the emergency procedures which will include shutting the process down. Also, they can handle normal malfunctions, although I like to include some challenging scenarios that are intended to get them to think a little deeper where they may not be successful initially.

I used to wonder if I was being too devious until I realized the real process sometimes delivered some very challenging problems to our Operators. The sign of a challenging scenario is when it will cause the first alarm to come in at a different system than the system with the problem.

Also, include at least one process leak scenario that requires them to isolate a part of the process and requires them shut down the plant. There is typically a lot of hesitation to do this. Most of the training is focused on keeping the process online. Having to make this decision at least once in training will help the trainee work through this if a large leak ever happens.

It is easy to be critical during simulator training because of the complexity of our processes and the fact there is so much going on. Remember to acknowledge when something is noticed or dealt with correctly. The intent is to help them develop a thought process.

It may seem helpful to point out every mistake, but it is better to help them understand how they arrived at their conclusions and to ask if there may be other information that would be useful. (I must remind myself of this last paragraph frequently).

When they are missing something, become curious about it. Explore their understanding and how they may have missed the important information.

## Case Study: Instrument Air Problems

In this example the conditions will be given and the steps that were taken during the incident. You will then have an opportunity to try to solve the cause.

**Background**

- The plant is old, built in the 1960s. Over the years there has been a few retrofit changes.
- 10+ years prior, each operations buildings' separate control room were merged into one centralized control room. The individual control rooms in the process units still have a control room board with gauges and alarm panels, but there is no longer control or the ability to control the process from the old panels.
- Over the years the instrument air system has become taxed from the addition of new equipment and old leaky fittings.
- Two years ago, the plant re-purposed a vessel as an instrument air receiver to increase the volume of the system and minimize pressure swings.
- When an alarm comes in on the boiler house panel, an alarm horn can be heard inside the satellite control room and the boiler house--it's a loud

air horn that can be heard over the noisy process. Another alarm will ring on the centralized control room DCS panel.

- The compressor discharge pressure runs at about 810 KPa. The instrument air header pressure looked good about 720 KPa, well in the normal range. The header pressure is read off a gauge in the boiler house satellite control room. The satellite control room also has an alarm panel with a low instrument air pressure hard wired alarm. Both the gauge and the alarm are in working order.

**The Scenario:**

Over the last week, a new additional instrument air compressor was added to the system. There was no room in the location where the old compressor was, so it was built into the system in another building, the boiler house/sulphur plant.

Today the compressor was started for the first time. The tie-ins and maintenance work were completed so the operations group decided to run the compressor to prove it out.

After proving out the compressor over a couple of hours running it, the next step is to isolate the boiler house from the rest of the instrument air in the plant to see the effect on the old compressor and the rest of the plant. The Operator closes the main isolation valve into the boiler house. He then goes back into the boiler house control room and watches the header pressure on the gauge on the control panel for the next hour, the gauge remains steady around 720 KPa.

The Compressor Operator reports the on/off cycle time of the old instrument air compressor has decreased, indicating that the compressor does not have to make up air as often, and therefore the load on the compressor has been reduced, just as they were hoping to see.

An hour later the Main Control Room Operator reports that he seems to be losing control of some of his control valves in the sulphur plant and boiler house. Several Operators head to that building to assist and all congregate in the boiler house control room. The instrument air header pressure gauge indicates normal as before, there is no hardwired alarm on the panel ringing in. A couple of Operators run down to the new instrument air compressor, the discharge pressure is still above 800 KPa. One Operator checks the discharge block valve and verifies that it is open.

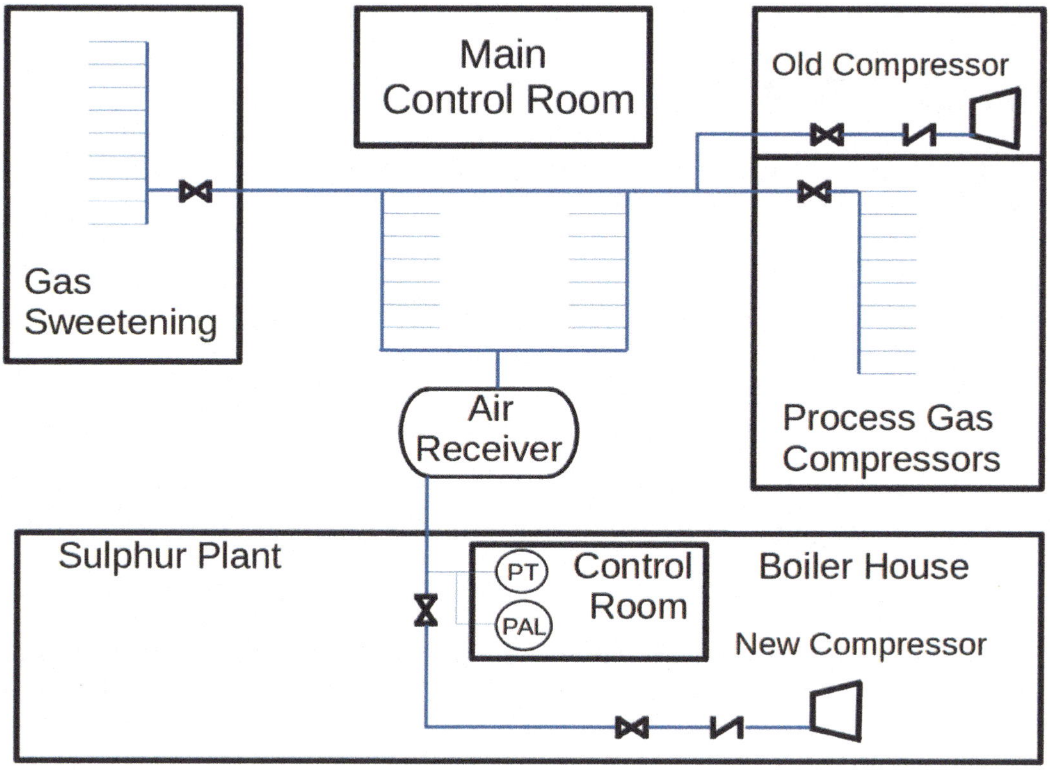


**The Response**

An Operator goes to the boiler house/sulphur plant isolation valve, climbing up a ladder to access it.

Opening the valve slowly the Operator notices a rush of air flowing through the valve as it is cracked off its seat. The Operator continues to open the valve and an alarm horn rings. The Operator also notices a 3/8 stainless tubing line, mostly hidden from view on the upstream side of the valve (not the boiler house side) tied into the valve flange, it is most likely a sensing line.

When the Operator goes back into the boiler house control room the instrument air pressure gauge is showing 120 KPa and rising, the low instrument air pressure alarm window is indicating alarm, and the Operators are confused and asking what you have just done. You explain the sensing line error and the rushing air at the valve.

What is the cause of the problem?

**The Cause**

There is an incorrectly (backwards) installed check valve on the new compressor.

***Contributed by Cliff Roy***

# CHAPTER 6
# Team Troubleshooting

*"Where all think alike, no one thinks very much"*
Walter Lippmann, 2-time Pulitzer Prize winner

## Responsibilities

Working together on troubleshooting efforts is critical in managing upsets quickly and effectively. Small incidents can evolve into large ones if the team cannot work together. Written and verbal communication skills are essential within the team and between the other shifts.

We will look at some best practices by area of responsibility.

**Shift Lead** (may also be called Foreman, Supervisor or Technical Leader)

- Assesses the larger picture and avoids getting drawn down into the details for any length of time
- Determines if the plant should be taken to a safe mode
- Acts as a focal point for other groups that may be required to assist in the efforts to solve a problem
- Delegates the action items that are required to validate theories and repair problems
- When there is time, they can ensure the documentation is completed for longer troubleshooting efforts
- It is important to keep your Shift Lead informed of the plant status (i.e. leaks, fires, rate changes, equipment outages, and process problems) to allow them to make effective decisions

**Panel Operator**

- Stabilizes process where possible
- Interprets process data used in the troubleshooting process
- Coordinates with the other Panel Operator if they are fortunate enough to have a backup
- Less experienced Field Operators will need specific instructions, e.g., check valve position, can you hear flow, check level gauge

- If you do not have a Shift Lead the Panel Operator typically takes on the responsibilities of the Shift Lead

**Field Operators**
- Field Operators are the eyes and ears in the field. Look for problems without having to be asked.
- Echo communication (i.e., repeat equipment or transmitter numbers to ensure you are working on the right equipment)
- Thorough rounds and checks will allow you to find problems early on

There are some behaviours that are detrimental to a team being able to function that are worth reviewing. These are likely symptoms of problems that have been around for some time. It would be worthwhile to reflect on whether you may default to any of these behaviors. During crisis situations, we can either shine or devolve by way of the following actions:
- Taking pleasure in other's mistakes
- Withholding information
- Creating an atmosphere of intimidation (i.e., behaving loudly and aggressively). This will shut down communication and ideas
- Rationalizing bad decisions. This shows up when the problem continues to deteriorate even when corrective actions have been implemented
- Group Think - We do not like to go against the consensus of the group. Ensuring everyone shares their opinions can prevent this
- Discounting other's opinions i.e., inexperience of others, other disciplines
- Shaming someone can cause their brains to revert to a primitive response impairing their cognitive functioning

The best way to work together as a team is to train as a team. Review emergency procedures as a team and have each member explain what steps they are required to perform in each scenario.

Develop scenario-based training exercises that involve the whole team. This is a great way for shared experiences to come out. You may also be surprised by the insights from the newer Operators.

These exercises will lead to an understanding of the teams, the individuals, and their respective strengths and weaknesses. You could call this the development of a team mental model which will help the team manage critical situations more effectively.

## Deeper Dive - Talking to Crazy

In the book "Talking to Crazy, How to Deal with the Irrational and Impossible People in Your Life" [25]Mark Goulston discusses the three streaks we commonly use to deal with situations when we feel we are being attacked or judged. Initially, we will revert to the primitive parts of the brain that seeks to protect us with our unforgiving streak where we will never let this person do this to us again.

If we hang onto this long enough it may generate a retaliatory streak where we may inflict the same damage onto the perpetrators. This can be emotional or social retribution.

If we continue along this path we will eventually get into estrangement where we will completely ignore the other person and not acknowledge they are even in the room.

You can imagine if we allow ourselves to go down this rabbit hole how ineffectively we will work together. Not only would this create an environment that would be very uncomfortable to work in, but eventually you will dread going to work. Our processes are far too complex to manage by ourselves at times and we need the support of others.

Becoming aware of these tendencies can help us to seek other ways to deal with personal differences. If you would like to learn more, I would encourage you to read the book.

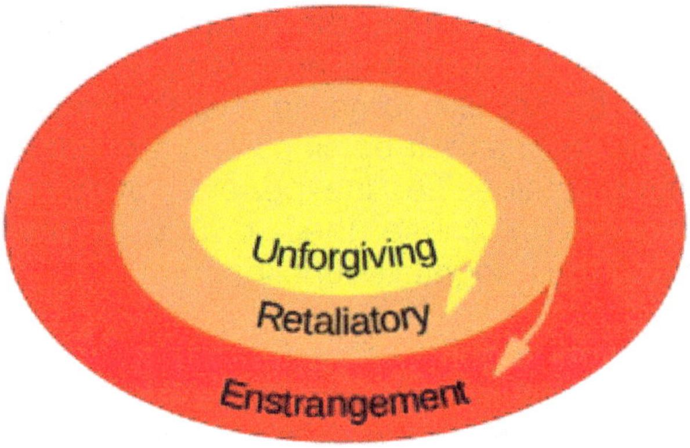

# CHAPTER 7
# Troubleshooting Tools

## Instrumentation Basics

Understanding some of the key features of your control system can save you some grief, and as a start, the best approach to troubleshooting is to stay out of trouble.

These settings may be somewhat different depending on the control system in your plant or have different terms to describe them. I am familiar with the Honeywell system. I recommend referring to the documentation for your system.

I hope to make some of the concepts clear as I have found some of the technical manuals difficult to translate.

I will start with the most common controller which is the proportional-integral-derivative or PID system. A PID controller continuously calculates an error value as the difference between the desired setpoint (SP) and a measured process variable (PV) and applies a correction based on proportional, integral, and derivative terms (denoted P, I, and D respectively).

You may not have to tune controllers as an Operator; however, it is helpful to understand why some valves respond much quicker than others when you make a setpoint change. This can have large consequences depending on the process.

Depending on the configuration, placing a controller in manual before a setpoint change can reduce the control action. Another option is to use setpoint ramps.

### Proportional

In the example below of a proportional level control system, the flow of supply water into the tank is controlled to maintain the tank water level. A fulcrum and lever assembly is used as the proportional controller. A float chamber is the level measuring element, and a 6-inch stroke valve is the final control element.

The fulcrum point is set such that a level change of 6-inches causes a full 6-inch stroke of the valve. Therefore, a 100% change in the controller output equals 6-inches.

This controller has a proportional band of 100% or a gain of 1. The fulcrum point could be moved so a 6-inch change in level would only move the valve 3 inch. This would be a proportional band of 50% (gain of .5).

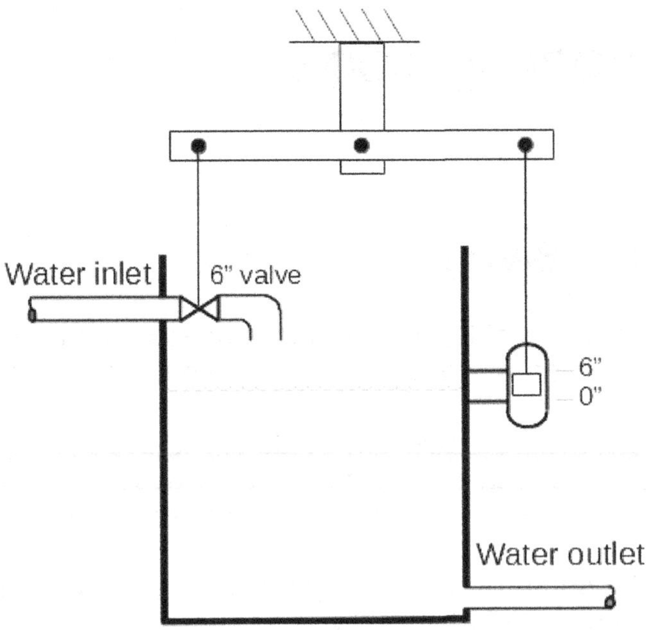


**Integral**

A proportional system will generally leave an offset of the PV to the SP as it is reactive to a change. Integral is used to check for offset and then increase the control action based on the time it has persisted. If the integral is 1 (minute) then the controller would implement the gain every minute based on the offset between the PV and SP until there is no error. The above example could not include integral. Another device to move the fulcrum point would be required to accomplish this.

**Derivative**

Derivative attempts to limit the deviation from setpoint by increasing the control action based on how fast the process variable changes per unit of time. For example, if you make a setpoint change on a temperature controller the output from the controller will increase more than it would have just by gain alone.

On the reverse of that, if the temperature is quickly reaching the setpoint the derivative will reduce the gain action to prevent overshooting.

Derivative may accentuate swings in the process on transmitters with a lot of noise such as flow transmitters.

## Initialization

In this example, TC2115 is the master and FC2151 is the slave with its setpoint normally following the output of TC2115.

When the Secondary controller is not in CAS the Primary controller's output follows the Setpoint of the Secondary controller's Setpoint. In the example below FC2151 has a range of 14. With a setpoint of 8, the output of TC2115 is calculated as 8/14 = 57.1. This will help to provide a bumpless transfer when the cascade loop is reconnected.

## PV Tracking

This feature can be set up on a controller in a cascade loop or just for a single controller. In this example when TC2115 is initialized from FC2151, or if it is placed in manual, the setpoint will follow the changes in the process variable while in either of these modes.

This feature is also intended to provide a bumpless transfer, although not all controllers have this feature turned on. It will depend on the control philosophy in your unit.

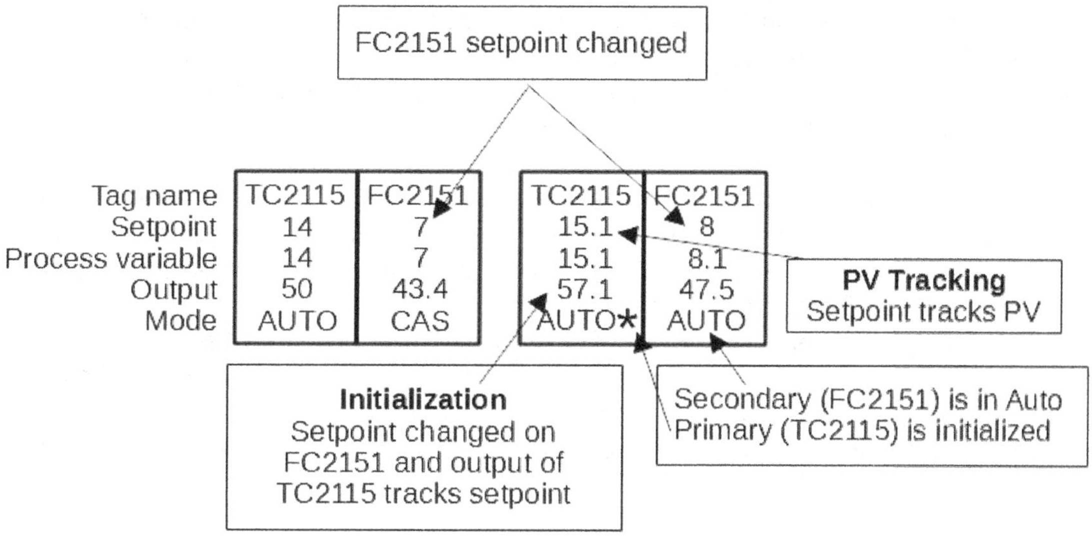


## Output indication

In the controller above (FC2151) the output indication is the output the controller is sending to the controller in the field. It is not feedback from the valve in the field.

If FC2151 were to fail closed in the field the output shown on the graphic would eventually ramp up to 100%. Some systems do bring in the feedback from the valve to the control room graphics. It is important to know your system to help with troubleshooting problems.

**Setpoint changes on a controller in Manual vs. Auto**

There is a parameter in a controller that determines whether a setpoint change will give it an extra kick on the Gain. This is called the Control Equation in the Honeywell system.

The tuning on a controller will also affect this. If the controller does not have PV tracking, placing the controller in manual before making the setpoint change will bypass this feature. Ramping the setpoint is another way to reduce the impact.

The intent of this control feature is to move the process closer to the new setpoint faster. This may or may not be beneficial depending on the system.

## Common Failures

Below are some examples of common failures. These can be expanded depending on the equipment in your process.

The more you think through and document all the possible failure modes, the more neural patterns you will have developed, and there will then be less cognitive load required to recognize and manage the problem.

My recommendation would be to start a database that is accessible to operations and all the other disciplines that can be built upon after failures happen. This would save time instead of individuals creating their own. It would also help to create a learning environment where everyone can share their experiences. Ideally, this would be in a database that makes it easy to categorize the different types of problems.

This chapter is not intended to provide all the possible failures that can occur. For more in-depth examples refer to "Troubleshooting Process Operations" by Norman Lieberman, or "Distillation Troubleshooting" by Henry Kister. Both these books go in-depth in design and systemic problems with specific examples.

**Instrumentation**

Based on my experience instrumentation failures seem to be the most common problem. Constantly cycling valves can cause premature wear. Sensing lines can be fouled by the process. The example below is for a basic flow controller, and the table provides some of the examples that can occur.

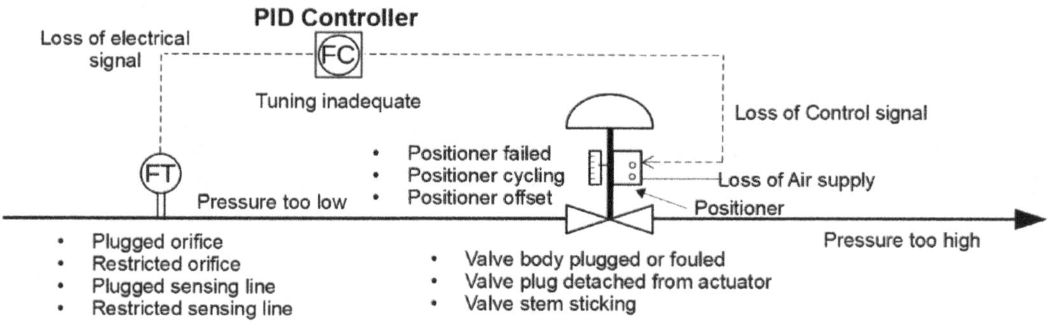


| Instrumentation Failure examples | | |
|---|---|---|
| **Problem** | **Signs** | **Possible Cause** |
| Transmitter deviation | Indication cycling or does not match normal flow for current output | Plugging or freezing of sensing lines, ambient temperature changes, bypass open on the transmitter, Improper calibration, |
| Sticking valves | Control erratic and intermittent | Gummed valve stem, packing tight, valve plug binding or plugged, diaphragm leaking |
| Cycling control | Output and transmitter cycling | Poor tuning, upstream or downstream process cycling, positioner failing |
| Positioner failing | Valve output could be off or fail fully closed or fully open or intermittent, valve position does not match DCS output, | Mechanical failure or wear, low air or loss of air, check positioner pressure gauges, Air supply pressure? Air leaks |
| Valve not responding to the output request | Valve at 0 or 100% | Is there an interlock tripped that is affecting the valve? Is there DCS Logic affecting the valve? loss of instrument air. |
| Loss of fill in transmitter leg | Indication will likely be reading higher | The leg was improperly blown down, process spike, wrong fluid, contamination. |
| Control valve does not travel completely | Valve travel does not match DCS output | Positioner out of calibration, Stem binding, valve plug fouled, foreign object in valve body |
| Valve passing at 0% | Flow when the valve is closed | Zero drift on positioner, reduce output below 0, ice on packing preventing full travel, worn plug, foreign object in valve body |
| Output card freezes | All control of valves on that card freezes | Control valve wiring removed without point being made inactive |
| Restricted flow | Valve 100% open and setpoint not being met | Fouled valve, debris in valve body, loss of differential across valve |

## Control valve failures

When the electrical signal is lost, or an Instrument Technician must remove the wire to the positioner the valve should be made inactive. The control hardware/software that provides the brains for a controller utilizes output cards to communicate with the control valves in the field. These are usually redundant. They likely provide communication to several valves. If the output card senses a loss of communication to a valve it will switch to the redundant output card.

If the loss of communication is because of the loss of electrical communication to one valve the output cards may switch back and forth a few times and then lock all the controllers in the last position, (hopefully) unless the valve is made inactive from the graphic. Be sure it is the valve and not the actual controller. Using FC2151, the valve tag name will be FV2151. Again, find out the details for the control system you are using.

## Pumps

There are many types of pumps, and all can have unique failures. Below are some of the common failures for centrifugal pumps that can be expanded on depending on the pumps used in your process.

| Pump Failure examples | | |
|---|---|---|
| **Problem** | **Signs** | **Possible Cause** |
| Cavitation | Discharge pressure low or cycling, flow erratic or low | Plugged suction screen, fouled suction piping, liquid too hot and flashing across impeller, suction or discharge pressure dropped, low suction head (level) |
| Vapour locked | Low discharge pressure or no flow when the pump is started | Vapours in pump casing, minimum flow line blocked in or restricted |
| Vibration | May be noticed by touch or noise, can also generate heat | Low oil, failing bearing, fouled impeller |
| Seal leaks | Visible to atmosphere or leak detection instruments | Seal fluid low, seal purge flow low, seal failing |
| Reverse rotation | Visual on motor fan blades, low discharge pressure | Electrical leads reversed on motor |
| Discharge pressure low | Pressure gauge reading low, flow indication low, levels not dropping | Valve closed, blind left in, relieve valve passing, impeller worn of fouled |

## Exchangers

Every process will require some form of heat transfer. The most common type of exchanger is the shell and tube. Plate exchangers are becoming more widely used as facilities are being designed larger. Plate exchangers have a larger heat transfer surface area in a smaller footprint. Because of the small clearances they tend to foul and plug off faster, sometimes even with filtration.

| Exchanger Failure examples | | |
|---|---|---|
| **Problem** | **Signs** | **Possible Cause** |
| Cross-contamination | Pressure increases on the low-pressure side, frost, analyzers, vapours on the low-pressure side | Tube leak, gasket failure, hot process flow on tube side, and no shell inventory can cause tube end seal to fail |
| Fouling or plugging | Outlet temperature changes, delta pressure high, | Contaminants in process, overheating, oxygen in the process, hydrates, construction debris |
| Loss of heat exchange | Outlet temperature changes | Vapours in process, approach temperature changes, composition changes, capacity exceeded, loss of liquid seal, |

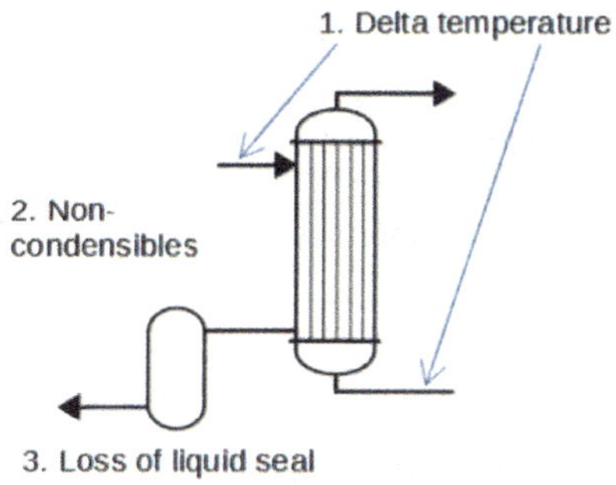


1. If the differential temperature between the heating medium and the process fluid is too low vaporization will be lost.

2. There are unwanted components that are unable to condense in the process conditions i.e. nitrogen in steam. They eventually fill the exchanger preventing the process gases from entering.

3. Latent heat exchange will be lost if the condensate pot loses all of the liquid and vapours blow through.

## Distillation

This is a very small sample of possible distillation problems. The concept of mass balance can be helpful during process upsets or large feed changes. Simply put, what goes into the tower must go out. This can help anticipate the tower's reflux, reboiler, and product draw.

Predicting these changes as the feed changes will help to keep the products on specification and minimize the product flow swings which may impact other systems or towers. Changes in the delta pressure and level changes in the bottom of the tower will also affect the mass balance.

Computer control systems will optimize the amount of product by allowing as much off-specification material as possible to go with the product. During process upsets, it may be prudent to run further away from the specification to give a little room for process swings. This may mean increasing reflux or bottoms temperature depending on your process.

Two concepts that help to understand distillation are mole fraction and partial pressure.

The mole fraction is the percentage of molecules of each component of the total. The mole fraction also determines the partial pressure of each component. The total of the partial pressures equals the vapor pressure of the mixture, and this determines the temperature of the mixture.

I have seen on several occasions where Operators (myself included) have overheated the bottom of a tower during start-up conditions and wondered why it took so long for the temperature to drop. The tray temperatures will not change until the composition in that tray has changed. If all the light ends have been driven up the tower, then they will have to travel down all the trays to the tray that is used for temperature control before its temperature will change. There can be quite a delay, especially during start-ups when there is little flow.

| Distillation Failure examples | | |
|---|---|---|
| **Problem** | **Signs** | **Possible Cause** |
| Tray flooding | Hi delta pressure, products off-specification | Foaming, tray fouling, improper tray installation, damaged trays or associated parts, excess feed rates |
| Overhead product off-specification | Analyzer, product temperature | Insufficient reflux, feed composition change, tray flooding, faulty analyzer, tower pressure dropped |
| Bottom product off-specification | Analyzer, product temperature | Insufficient heat – reboiler malfunction or capacity exceeded, feed composition change |
| High tower pressure | Pressure indicator, reboiler capacity maxed, | Feed contaminated with a lighter component, condenser malfunction or capacity exceeded |

# Unit Awareness

Maintaining an awareness of the unit throughout the shift allows you to be more pre-emptive than reactive to problems. Completing a thorough review of all the graphics at the beginning of the shift will help pick up problems before they escalate and help with situational awareness if an event happens.

Being able to pick up on an anomaly is dependent on being familiar with what is normal. Being able to use our pattern recognition skills requires a pattern in which to compare a change. One tool to develop this is a log sheet to write down the setpoints, process variables and outputs (see below for an example).

The intent of this is not necessarily to be able to recite what every value is, although eventually, we will be able to consciously recall the key parameters. Documenting values like this helps us to focus on the unit and eventually allows us to develop patterns to draw on.

A common term used is rounds and checks. It can start with filling in a log sheet. Rounds and checks should also involve viewing the graphics with a curious mindset questioning the things that catch your eye.

This is called "Staying Out of Trouble" as you notice problems before they escalate.

**Rounds and checks checklist:**

- Alarm summary, disabled alarms
- Controllers in correct mode and advanced controllers on
- Product specifications within range
- Valves or process variables cycling. This could be an indication of a problem forming with the valve or the transmitter
- Valves that are 100% open or closed
- A process variable that has a very flat trend could indicate it is at the top of the range of the transmitter

- Look for system errors of the control system
- Read the logbook. There may be information the support staff may have entered and or that your relief forgot to mention.

| LOG SHEET | | | | | | | | | |
|---|---|---|---|---|---|---|---|---|---|
| | | | 11/4 | 11/5 | 11/6 | 11/7 | 11/8 | 11/9 | 11/10 |
| | Plant rates | PV | | | | | | | |
| T201 | FC2001 | SP | | | | | | | |
| | | OP | | | | | | | |
| | PDI2017 | PV | | | | | | | |
| | TC2015 | SP | | | | | | | |
| | FC2051 | SP | | | | | | | |
| | | OP | | | | | | | |
| | LC2010 | SP | | | | | | | |
| | FC2021 | SP | | | | | | | |
| | | OP | | | | | | | |
| | FI2017 | PV | | | | | | | |
| T211 | FC2101 | SP | | | | | | | |
| | | OP | | | | | | | |
| | PDI2117 | PV | | | | | | | |
| | TC2115 | SP | | | | | | | |
| | FC2151 | SP | | | | | | | |
| | | OP | | | | | | | |
| | LC2110 | SP | | | | | | | |
| | FC2121 | SP | | | | | | | |
| | | OP | | | | | | | |
| | PC2113 | SP | | | | | | | |
| | | OP | | | | | | | |
| | LC2120 | SP | | | | | | | |
| | | OP | | | | | | | |
| | LC2130 | SP | | | | | | | |
| | FC2141 | SP | | | | | | | |
| | | OP | | | | | | | |
| | LC2110 | SP | | | | | | | |
| | FC2121 | SP | | | | | | | |
| | | OP | | | | | | | |

## Trends

Process trends are one of the most powerful tools in troubleshooting problems. They help to level the playing field when it comes to how much data is provided to a Panel Operator. It is well worth the time to get to know all the features of the system you are working with.

In the trend below, the SP (Setpoint) and the Mode have not changed. In each system there will be a number associated with each mode, e.g., 1 may indicate that the controller is in Auto. In this example the vertical numbers on the right represent the output of the controller or the process variable of the point was selected.

To the left of the vertical blue line the PV (Process Variable) is moving around normally.

To the right of the vertical line the PV is trending down and then the output starts to increase as the controller sees the deviation between the SP and PV. The rate of change of the Output will depend on the controllers tuning as the controller is still in Auto.

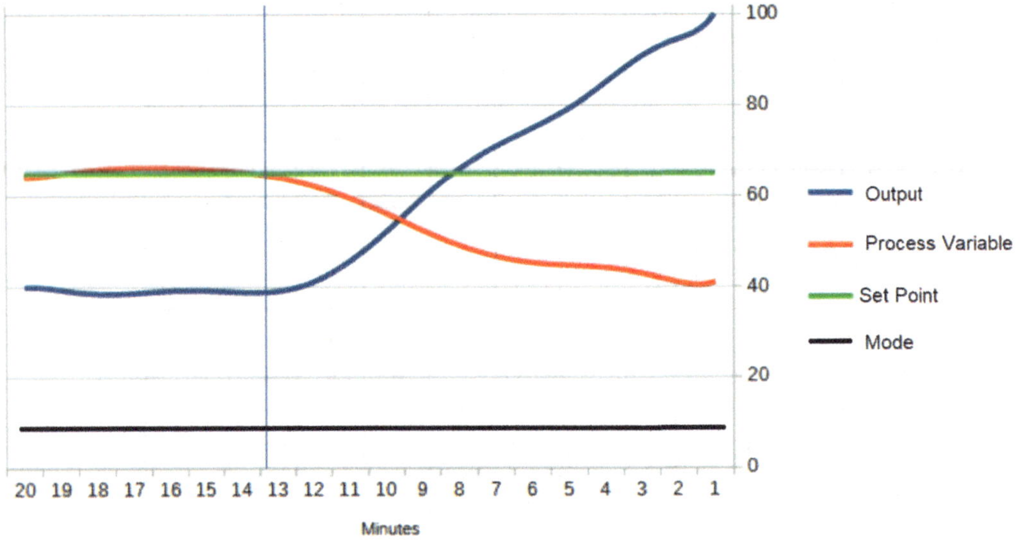


If you arrive part-way through an incident that others are struggling with you can develop some trends of the key points after you have been given an overview of the problem. Trends are helpful to cut through the assumptions that may have been made.

**Trend Limitations**

The historian that I am familiar with saves data for each point once per minute except for a few points that are more frequent depending how critical they are.

If the point is already showing on the trend, it will update every few seconds and the trend will have much more resolution than if you just called up the point. This may only be available on the trends on your DCS or control system graphics. If you are viewing data on the other side of the firewall to the historian, the real time updates will not be available.

If you are calling up the points of two variables that changed within seconds of each other it may be difficult to see which happened first depending on where the data was collected. A lot can happen in one minute. Even a spike from a transmitter can be lost in a one-minute average.

# Alarms

Alarms can either be an Operator's best friend or their worst enemy. They are an excellent warning system when the process is exceeding parameters. Unfortunately, due to the number of data points and possible alarms that can be assigned, alarms can become a distraction.

**Alarm Rationalization**

Alarm rationalization is a common practice in the industry. However, this is typically viewed from a steady-state operation to give the Panel Operator enough information to prevent an outage or escalation of a problem. The intent is to avoid repetition of alarms to allow the Panel Operator to quickly sort through them and determine the cause of the upset. This can be very effective when there is a deviation of a transmitter, failure of a control valve, or loss of a pump.

With only two to three alarms the Operator can zero in on a problem and initiate a response and minimize the upset. It already has been covered that an alarm can interrupt your train of thought for up to 40 seconds.

Alarm rationalizations are effective for smaller events. During a large event, like a plant outage, the number of alarms can exceed 100 even with alarm rationalization.

During these events, operations are required to safely isolate and secure the process. This task can take some time limiting the Panel Operator's ability to address the alarm flood that occurs. An alarm notifying the Panel Operator of an event that develops during the outage can easily be missed, allowing the event to escalate. This could be a control valve passing that has been closed or a process loss of containment.

This type of event occurred in a petrochemical facility during a complete outage when a pressure control valve to flare was passing and sub-cooled a distillation tower. The low-pressure alarm was missed in the several pages of alarms. The additional flaring was hidden with the normal flaring required during an outage.

**Smart Alarming**

Programs can be developed to reduce the number of alarms during large events to a manageable level for the Panel Operator. A program can be used to disable or change the trip point of an alarm. A hazard assessment of which alarms are known to come in during the specific event can determine which alarms are expected and will not cause an additional hazard if they are disabled or altered.

If the number of alarms is reduced to less than five, the Panel Operator can quickly scan the alarm summary to look for anomalies that are not normal for that type of outage. Detection and identification of problems, particularly independent failures, are improved with less visual and audible noise in the system.

A study conducted for the Center for Operator Performance demonstrated the value of “Smart Alarming”.[26]

**Hardwired Alarms**

These could also be called lightboxes or other terms depending on your facility. They are fixed in one location to make them easily noticed.

Depending on the design philosophy, they typically are emergency notifications when a major part of the process is close to tripping or has tripped. They are also used for safety systems to alert operations for gas detectors, fire detectors, or safety showers.

Review each alarm to understand what signal is initiating the alarm and what the trip points are. Some of the signals may come from switches in the field (e.g., level, pressure) or transmitters. Understanding the source will also help to troubleshoot if there is a failure.

Planning for potential responses will relieve stress and increase your response time. Review which graphics you will need to validate the alarm and which graphic you will need for response to the alarm.

## Case Study: De-Ethanizer Hydrates

**Hydrates restricting reflux flow after a petrochemical facility startup.**

After the start-up of a petrochemical facility, the feed rates were increased over the evening. The normal reflux rates on the de-ethanizer tower could not be reached which limited the rates through the entire process.

The de-ethanizer typically runs at 2300 KPa (330 PSI) with an overhead temperature of -12 Celsius (10 F).

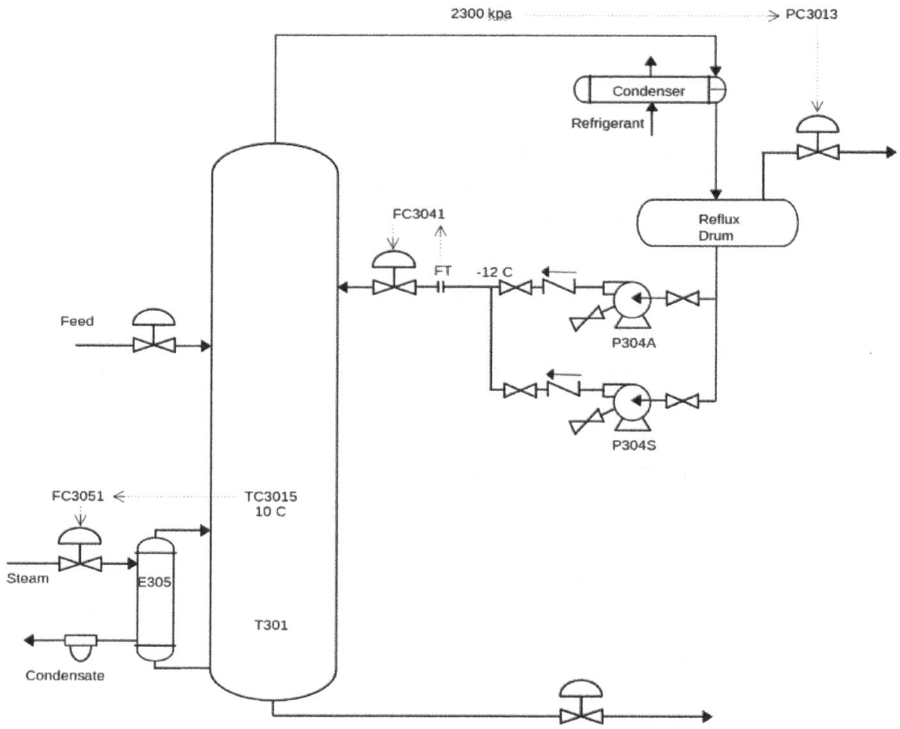


During the night shift informal troubleshooting determined hydrates as a plausible culprit and pumped propanol through the reflux control valve and bypass. This resulted in slight improvement in reflux flow but this was short-lived. The reflux valve was injected with propanol several times throughout the last few hours of the shift, but not enough recovery was made to continue with putting the tower fully online.

The day shift pulled together a quick troubleshooting team and made a drawing of the process on a white board. A brainstorming and troubleshooting session focused on the flow orifice or the pump check valve being the source of the restriction, with potential for hydrates as the main problem. Part of the team went looking for orifice engineering data and piping spec information, part of the team started a more formal troubleshooting effort, and the other half of the team proceeded to the field to find points to inject N-propanol upstream of the flow orifice.

The field team determined that the only point to inject propanol was on the pumps casing drains. One of the pumps secondary mechanical seals failed at this time. It was noted that the running pump discharge pressure was slightly less than normal and fluctuated slightly.

The focus then moved to a possible restriction in the pump suction. While scoping this out it was noted that there were strainers located in the pump suctions. The strainers were noticed by the metal tabs that stuck out of the flange on the pump suctions.

## The Solution

There is no easily assessable injection point upstream of the reflux pump suction strainers. With help of a Process Engineering calculation, it was determined that if we pumped in approx. 1.5 barrels into the suction of the offline pump the propanol would spill over into the suction of the online pump putting propanol in upstream of the suction strainers.

The offline reflux pump was isolated at the discharge and N-propanol was injected into the pump casing drain which eventually back filled into the other pump and removed the hydrates that were in the suction screen of the online pump. The pump discharge pressure increased, and the sound of the pump changed. There was an increase in flow to the tower and the overhead temperatures started to drop.

Feed through the plant was allowed to continue now that the reflux flow and control valve position were normal.

The moisture to form the hydrates was likely left over in the de-ethanizer feed exchangers and possibly the de-ethanizer tower trays. They had taken an extended period to reach acceptable dew points during plant recommissioning activities. This moisture was likely driven overhead as the tower was heated up to normal operating temperature.

## Background

After a process facility has been removed from service for maintenance and cleaning there can be moisture left behind from cleaning the towers or exchangers. This can include chemical washing or hydro-blasting. Pressure testing equipment using hydro-tests with water is also a common practice to check the integrity of the equipment, including welds.

In processes where there is a potential for hydrates to form, the moisture must be removed. Pressure drops across valves or plate exchangers will lower the process temperature and increase the likelihood of hydrates.

After the free water has been drained, nitrogen is typically used to purge out the remaining moisture. This has the added benefit of removing the oxygen from processes with hydrocarbons or reactive components. The dew point is then checked using either an analyzer or the old school tester that flows the nitrogen past a mirrored cup with acetone cooled with dry ice and a thermometer.

The dew point temperature required will depend on the temperature that the process will operate at.

N-propanol has an affinity for water and has a normal freeze point of -129C making it an acceptable choice for removing hydrates. N=propanol can act as a poison in some systems and must be flared off downstream of the process injection points.

**Contributed by Ed Noble**

# CHAPTER 8

## Critical Decision Making

Everything discussed to this point is to prepare you for making critical decisions. Without thorough preparation our chances of success decline.

**Here are some highlights of the concepts that have been covered:**

- Understanding the core principles of our processes
- Knowing how we make decisions, like the Recognition Primed Decision model, combined with understanding how to develop patterns and insights
- The use of procedures for tasks with critical steps. They also help us to develop a pattern of the process that can be drawn upon during an event. Having a procedure to follow is the best type of situation to be in. The rest of the preparation is for those situations for which we do not have a procedure, or the problem does not exactly follow the procedure.
- Training is a key to prepare ourselves for critical decisions. Start with the basics and work up to understanding more complex patterns.
- Managing our stress and physiological responses will help us to stay focused and clear. This will enable us to move from instinct to insight. If you are stuck, accept that you do not know the answer but that you can find the answer. If you start to be concerned about what people are thinking or what they will say if you fail, just breathe and let the thoughts pass, and try to refocus on the task.

I do not believe there is one magic way to prepare for making critical decisions. I hope I have provided some helpful models and techniques that will help you in these circumstances.

Most critical decisions Operators must make are about keeping the process online and on specification. Sometimes it may involve responding to process leaks (loss of containment). Hopefully you never have to respond to a large fire or an explosion.

The intent of this book is to cover how to troubleshoot process problems. I cover some topics on preventing process safety incidents (leaks, fires & explosions) in Chapter 9. Another book should be written on how to respond to process safety incidents for our industry. There are several on the topic of designing process safety systems. I am not aware of any that address how to manage a situation when the systems fail.

# Prioritizing

There will be times when you will be overwhelmed by an event. Alarms, phone calls, radio communication or conversations in the control room can all be taxing to you. You will have to prioritize what will need attention the most or require immediate attention.

The **first** concern is **protecting life**. Clear anyone out of the area if they are in danger. Use the emergency system when needed to clear everyone from the unit. The system I am used to has an alert that signals everyone except operations to clear the unit when there is a large process upset or a leak. The evacuate signal is used to clear everyone from the unit.

The **second** priority is **process safety**. Is there a possibility that the maximum or minimum limits of the equipment could be reached? This can involve pressure, temperature or even differential temperature on some exchangers. Understand these limitations before the event or know where to quickly find this information.

The **third** priority will be **keeping the process online and on specification**. Typically, if this is achieved personnel and process safety are also addressed.

It has been said that we cannot consciously multitask. So how do our minds handle these complex situations, and how do we prioritize what must be handled first? Most of us have seen or heard of Operators that can look at a situation and almost immediately tell you what the problem is and how to do fix it. My ability to hone in on a problem has greatly improved after 18 years of simulator training and seeing thousands of upset conditions.

# Reaching The Zone

Operators may have months of not being challenged by a demanding event and then one happens suddenly without warning. They will be performing the day-to-day routines of rounds and checks, preparing equipment for a regen, or managing preventative maintenance being completed by Instrumentation, Millwrights, or Electricians. When a process upset occurs, the Operator's mental processing will have to go from 30 mph to 100 mph in seconds.

Athletes talk about being in The Zone and how everything just clicks for them. I have witnessed Operators in this state where they just seem to pick up everything and key into the source of the problem quickly and know how to respond.

In "The Art of Learning" Josh Waitzkin[27] discusses how he prepared for martial arts competitions. The first step is to break down the required steps into chunks. When we have thoroughly learned the basics, we can then move on to larger chunks and start connecting the chunks together. Chapter 5 on training describes how to break the process down into chunks. We must also do this for our responses. Even an emergency procedure with 100 steps can be broken down into chunks.

This strategy is what leads to pattern development. To use these patterns, it is best to be in a state of open curiosity to allow our subconscious mind to detect patterns. Our conscious mind may not be able to multi-task, but our unconscious mind can sift through large amounts of data and make sense of complex situations (if the preparation has been done).

When we are in The Zone, the conscious mind directs the focus of what is important and interprets what the subconscious mind identifies. This would be an alternating and shifting of awareness utilizing the pattern recognition of the unconscious mind to the focus and interpreting by the conscious mind. (This is covered in Chapter 1.)

If we have not developed pattern recognition, our conscious mind will become overloaded trying to sift through all the information. This will greatly increase our response time and likely lead to the event getting away from us. On the flip side, utilizing the unconscious mind to sift through the large amounts of data available frees up the conscious mind to make alterations or corrections as required.

If our training is thorough, our unconscious mind will be able to detect the patterns of the big picture while our conscious mind can focus on the task on hand.

When we are in The Zone, we can focus on the problem, and we are not concerned with what people are thinking of us personally or if we will have a job when this event is over. We are still open to comments and suggestions relating to the challenge at hand.

## Mastery

Aikido instructor George Leonard on mastery:

"How long will it take me to master Aikido?" a prospective student asks.
"How long do you expect to live?" is the only respectable response.

If you apply this quote to Process Operators, we will stop learning when we retire.

## Mastery is a Mindset

A mindset is the lens through which we filter and perceive life. If we have a growth mindset, we have a belief that our abilities can be developed. If we have a fixed mindset, you believe you are born with your abilities, and they cannot be changed.

In her book called "Mindset, The New Psychology of Success"[28] Dr. Carol Dweck discusses how a growth mindset (a belief your abilities can develop) or a fixed mindset (a belief you are born with your abilities, and they cannot be changed) can have a dramatic impact on our success.

A belief that the only way to improve is by eliminating errors, is a fixed mindset. The belief that performance improvements depend both on cutting errors and on making insights is a growth mindset.[29]

I do not believe that a mindset is something we can just turn on and you are done. It is more of a journey of becoming aware of how you perceive situations. A mindset is developed based on your experiences and can be heavily influenced by your family's mindset. As most of us do not have memory until after we are 4 or 5 years old this may be difficult to know what events influenced our mindset. This delves more into behavioural psychology, but I do believe keeping our mindset in "mind" can help us to gradually make changes to the lens through which we view events.

Something I have observed from master Troubleshooters is they are continually trying to understand how things work even when they are not broken. They keep exploring how something works until there are no questions left. They are never comfortable with a superficial explanation of something.

## Mastery is Pain

We could also think of this as agitation or strain. On the days that we are the least motivated, we must continue to strive for improvement. We need to summon up our curiosity to keep building expertise even if it is a small amount. It must become a habit. Whenever there is something we do not understand it has to be followed up on right to the conclusion. Our ability to respond to problems needs to become our professional persona.

This means forecasting and trying to figure out what will happen if... this valve fails, or a pump shuts down. See Chapter 5 on scenario-based training for examples of exercises that can help you move through those times that are difficult to find some motivation. I have received some very positive feedback on the pattern-developing exercises mentioned in this chapter.

In the Podcast "How Your Nervous System Works & Changes | Huberman Lab Podcast #1" on You Tube,[30] Dr. Andrew Huberman discusses how focus, agitation, and strain are required to trigger neuroplasticity. This is the term used when our neurons make new connections when we learn something. This reconnecting and strengthening of the neurons happens during sleep.

You may find that you have experiences that can be demotivating. I will not mention the various ways this can happen as you have probably experienced some already. The challenge is to eventually pick yourself up after these events and remember that you are doing this for yourself. The reward will be when you have been successful against all odds, and you know that you are becoming a master at what you do. You may even receive the occasional recognition to validate this when you become the go-to person for questions, or you are invited to participate in starting your facility up or become involved in projects. The truth is that Operators are terrible at giving positive feedback, myself included. You must have some faith that if you keep working at it you will get there. When I started operating I was told that it takes 10 "Atta boy's" for 1 ”ah shit”. I believe I am still behind.

Some believe it may take up to 10 years to master a profession. I believe that this time can be shortened with the suggestions and exercises mentioned in this book.

### Mastery is an Asymptote

A horizontal asymptote is a straight line that a curve approaches but is never reached. Mastery is something you can approach. You can home in on it. You can get close to it, but you can never reach it. This is very fortunate. You were likely worried you would get bored after becoming a master Operator. Our processes are usually so complex, we will even be surprised after 30 plus years in the business.

Use the sense of discovery as an allure to continue. The day you think you know it all will be the day your process will show you differently. This ties back into the concept of mindset. If you believe you know everything you have a fixed mindset, and when you view a problem this way you may miss what it is trying to show you. Of course, you will have events that you will figure out quickly and efficiently. Even these have something to teach you from some of the secondary impacts.[31]

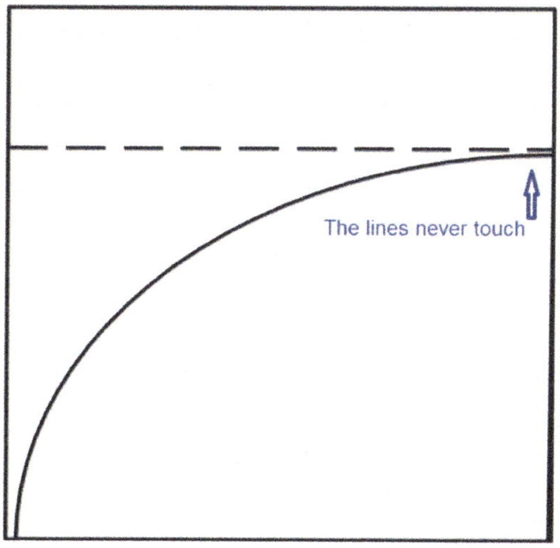

## Deeper Dive - Neuroplasticity

The way our neurons communicate with each other is with dendrites that are like branches from a tree that reach out to other neurons. Studies with animals and humans have shown that when they are in an enriched (challenging) environment they grow more lavishly.

The term neuroplasticity is becoming more common as researchers are finding that our brains are continually changing, even when we are older, although somewhat slower. You can teach an old dog new tricks; it just takes a little more effort (assuming they want to change).

The brain does this with neuromodulators. Acetylcholine is an important messenger that increases plasticity in targeted areas. The neurons that provide the acetylcholine must first be activated by relevance. This means if you are trying to learn something, and you just don't care, that area of the brain will not change its connections and you will not learn anything.

A formula is required for neurotransmitters to make changes in our neural network. These include emotions like curiosity, interest, and investment. These are emotions that may be hard to muster when you are a shift worker. One way to increase curiosity is to set the training up as questions and scenarios as described in Chapter 5 on training. Interest should stem from previous experience in how mechanical things function. This could be a challenge if it was only a paycheck that attracted you to the job. Your investment is not only in your reputation at work; rather, it is your understanding that you are an integral part of keeping your process running and that your actions could cause harm or even death to someone.

It can be a challenge to stay engaged and continually try to learn new things. The reliability of the process can have an impact on this. If the process is reliable and there is lots of automation the Operators may not be challenged that often. Except that during an upset they will have to go from being on watch to having to think and respond within seconds. In this case, it is very important to continually challenge yourself by asking questions and thinking through possible upsets. If neuron connections start to lose their relevance (not used), the connections may be lost.

If you would like to learn more about neuroplasticity, I would recommend "Livewired, The Inside Story of the Ever-Changing Brain" by David Eagleman.[32]

## Case Study: Turbo Expanders over-speed

This is an excellent example of when a Panel Operator had to make a critical decision. It involves a Turbo Expander in an ethylene manufacturing facility. The emergency block valve located on the inlet to the expander failed to close during a safety system shutdown. A high level in the expander suction drum initiated the shutdown. Refer to the diagram on the following page.

The shutdown logic included the following steps.

- Close the shutdown valve on the expander suction
- After 30 seconds close the compressor shutdown valve and the compressor false load valve. (This step is to allow some time for the compressor to slow down the machine)
- Close the suction pressure control valve if it is in Auto (This was to prevent the valve from winding open when the machine was tripped. It was not part of the safety logic but initiated by it)

This logic looks like an effective way to safely shutdown a rotating machine.

Let us look at the events that led to this machine exceeding the high-speed trip by 12,000 rpm. Refer to the diagram on the next page.

- A new shutdown valve was installed on the expander suction (the step to heat the internals of the valve was missed before the installation. This would have reduced the porosity of the metal and its ability to hold moisture)
- Three days before the event the machine was unstable, and the decision was made to not exceed 19,000 rpm. This was managed by placing the expander suction pressure controller in manual at 70%
- A large rate reduction was initiated to minimize flaring. This was required as the facility was being shut down due to an internal exchanger leak that was contaminating the product.
- Since the expander suction pressure controller was in manual the pressure in the DeMethanizer and Methane stripper dropped by 250 KPa (35 psi). This led to a large amount of vapor leaving the DeMethanizer overheads and filling up the reflux drum. The reflux drum then overflowed and filled the expander suction drum that initiated a high-level trip to the machine.
- Unfortunately, the expander suction shutdown valve did not close and the machined exceeded the high-speed shutdown.
- The panel operator closed the expander suction pressure controller after 50 seconds and stopped the machine

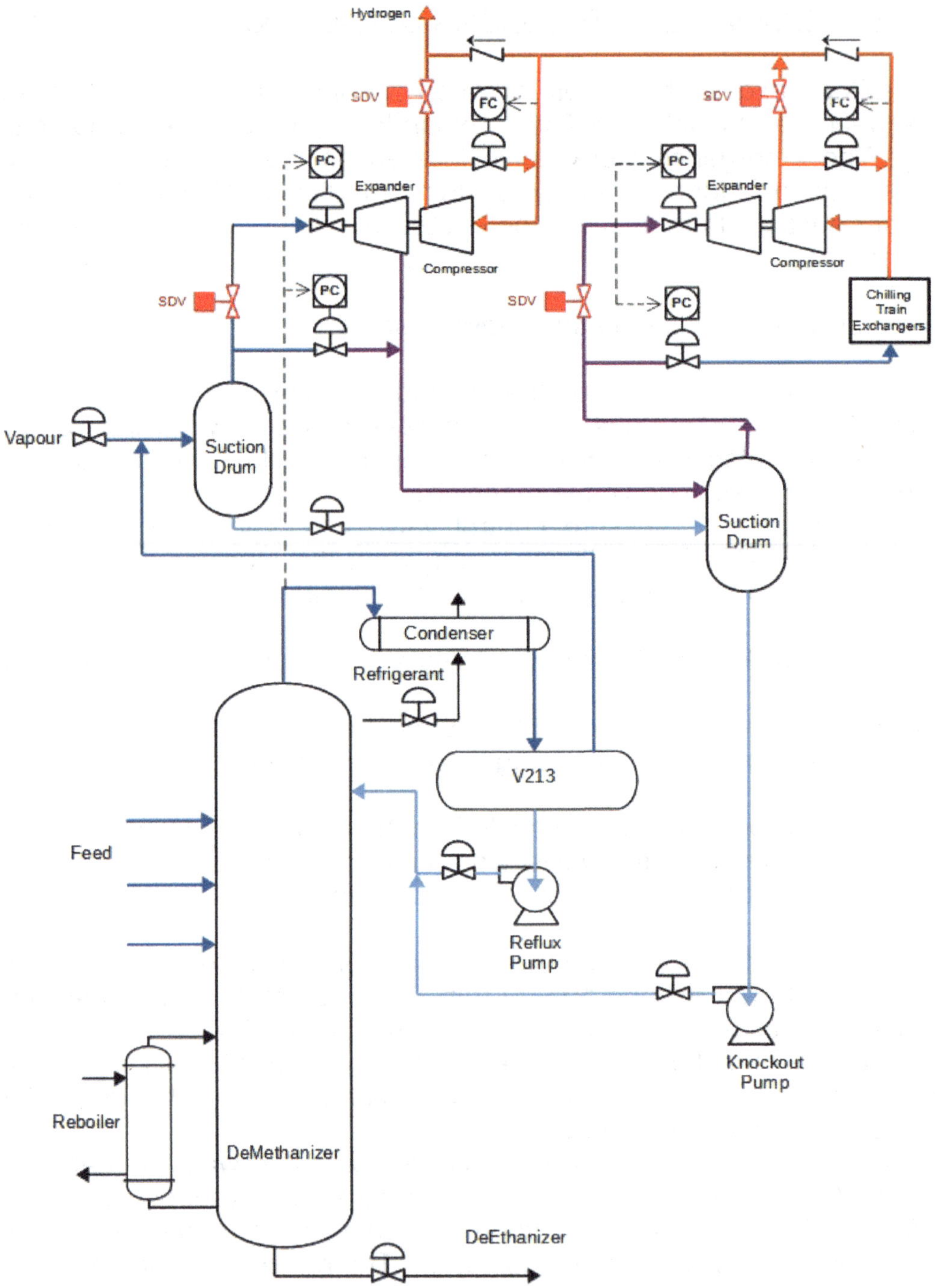


This is an example of an event where the quick response of the panel operator prevented the destruction of the machine. In hindsight we can see this was an obvious solution to solve the problem. It would be easy to overlook that they were also responding to a 50% reduction in feed that would have resulted in several pages of alarms and managing the impact of off specification product in the pipeline. There are many important process moves that would be required by the panel operator that will not be mentioned here.

Somewhere in the middle of all that the panel operator recognized the problem and closed the valve, within 50 seconds of the machine over-speeding. There was no procedure that addressed this issue. There was no documentation that suggested it as a possibility. There was a risk review about 25 years ago prior to the machine being put into service that would have included several experts determining what safety logic to use, with no time pressure.

The intent is not to criticize the individuals' decisions leading to this event, the intent is to recognize that our operators will have to interpret data and make decisions that have not been proceduralized or even thought of. There will be times when not every sequence of events can be anticipated. Which is why operations must be trained on how to sort through data, understand the critical data and how to move to a decision.

As a side note these are the changes that were made to the Turbo to prevent re-occurrence. The more experiences like these we share the safer our industry will be.

- Heat treated the expander shutdown valve internals to prevent ice formation
- Review the type of valve packing on the expander shutdown valve to ensure it is not impacted by the very cold service
- Delay the closing of the compressor shutdown valve and false load valve until the speed is below 5000 rpm. Flow through the compressor slows the machine down.
- Close the expander pressure control valve during a shutdown if it is in Auto or Manual

**Contributed by Author**

## Keys to Troubleshooting

Becoming an expert Troubleshooter requires effort. That's bad news if you are hoping for easy answers, but good news if you relish a challenge and an opportunity to gain elite status. You can wait for the school of hard knocks to teach you (not recommended as this usually results in loss of production, equipment damage or personal injury). Or you can prepare yourself in the following ways:

**Gain a thorough understanding of how the equipment functions and how the process works.** This can be thought of as First Principle thinking. A first principle is a basic assumption that cannot be deduced any further. It starts directly at the level of established laws of physics etc. This contrasts with analogous thinking which is comparing the problem to a past problem. These principles are the first step to building a mental model.

First Principal Examples:

- Thermodynamics – heat transfer, state changes
- Mass balance – what goes in must come out
- Chemical reactions – endothermic, exothermic
- Differential pressure

**Learn the facts.** Some must be memorized, and you will have to know where to readily access the rest. Modern control systems are making this more convenient where certain details can be called upon the display or links can be added to critical documents.

Past incidents are a great resource and reviewing the incident through your historian tool can fill in the blanks and details that are left out of the incident report.

Reviewing the Chemical Safety board videos is a must for every new Operator. Not only are the investigations very thorough, but the graphical videos they developed are also phenomenal. They are very helpful in piecing together the sequence of events and almost give you the feeling of being there. We are very fortunate to be able to learn from events like these and I am grateful to the authors and all those involved in producing them.

Another way to capture incidents is to include a section in the training manual dedicated to explaining some of the key learnings. This makes them available to a new Operator that will not likely go back and review every incident from the history of the facility. It can also be set up more like a training tool than just capturing the key points that are typically in an incident report.

Reviewing the incident with the Operators that were involved or an experienced Operator using the historian will flush out some additional learnings that may not have been captured from the incident report, including suggestions on how to avoid a similar incident in the future and how to manage the event if the situation repeats itself.

**Practice going through known scenarios.** This will reduce the cognitive load during an actual event and enable you to deal with all the variables much more effectively. It is not uncommon for a secondary problem to show up during an event. It is important to know how to respond to known situations. However, being able to figure out the cause of a problem and how to correct it is the true measure of a master Troubleshooter. Refer to the scenario-based training section in Chapter 5.

**Have a questioning attitude and actively look for problems.** Explore why something is different than before. Have a curious mindset while looking for insights. Learn to anticipate and consider what the secondary impacts will be from the upset. For example, if reflux is lost to a distillation tower, what will happen to the product quality, what will happen to the bottoms temperature?

Occasionally we need to shift our focus. We tend to focus on the details and lose sight of the big picture. Occasionally come up for air and see what else is being impacted. Not shifting our focus can also lead to fixating on one problem at the expense of the rest of the unit. When we are trying to make sense of a situation, we may stick with the first plausible cause even when other evidence is indicating otherwise.

When we have a problem, we may have a hunch but cannot be sure until we validate it. We must test our hypotheses. Have the Field Operator check local indicators, gages, valve outputs or blow down a transmitter.

Valves can be stroked to see the response of the process and validated in the field. The degree that the valve is stroked will depend on the process as you do not want to cause a secondary upset. There may be duplicate transmitters or by using other transmitters you can calculate a mass balance of the feed entering the system to the feed exiting the system.

Do the analyzers confirm what you are seeing, or the vice-versa, do the process variables confirm what the analyzers are indicating?

## Maintaining Knowledge

Maintaining our knowledge is also an important factor in keeping our troubleshooting skills sharp. This can be even more of a challenge if you do not frequently work that unit. Training others is a very helpful method.

Reviewing the key steps of emergency procedures every few weeks will help to reinforce the steps and reduce the likelihood we will freeze up during an event.

The diagram below shows the baseline as the state that we begin at. In the acquisition phase, we are learning how the process works. The extinction of this knowledge will happen if we do not review the information periodically. Our brains are efficient— if neural connections are not being used then they will be assigned to something else.

The maintenance phase is required to keep the knowledge and skills intact. To become proficient Troubleshooters, we also must continually grow-- not only to get a deeper understanding of the process but to keep on top of changes.

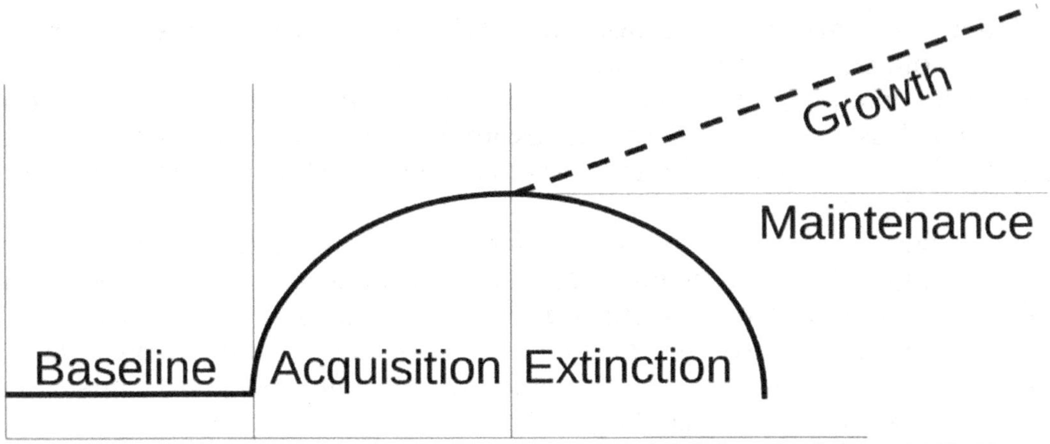


Adapted from "Enhancing Learning in Training and Adult Education". Ronald R. Morgan, Judith A. Ponticell, Edward E. Gordon[33]

And finally, here is the last key to troubleshooting.
**The 8 Ps** - **P**roper **p**rior **p**lanning and **p**reparation **p**revents **p**iss **p**oor **p**erformance

## Case Study: Polyethylene Level Transmitters

This example shows the importance of knowing the first principles.

A fluid bed polyethylene reactor requires several differential pressure and level indicators to ensure proper operation. A purge header is used to supply an ethylene purge to the transmitter tie points to prevent the polyethylene powder from plugging the nozzles. The reactor pressure was about 2000 KPa (290 psi) and the ethylene purge header pressure was 400 KPa (58 psi) higher.

During the first winter after start-up the transmitters indications were becoming erratic, particularly at the top of the reactor. After referring to a phase diagram for ethylene, at its current pressure, it was noted that ethylene would start to condense at the current ambient conditions.

The ethylene purge header started at the bottom of the reactor and extended up approximately 35 meters (115 feet). The line was uninsulated and exposed to the wind. Due to the small flow, the ethylene was cooled until it started to condense causing liquids to affect the transmitter sensing lines.

The issue was resolved by insulating the purge header.

***Contributed by Tracy Walsh***

# CHAPTER 9
# Extras

In this last chapter, I want to highlight that the following topics can contribute to ensuring the safe operation of your facility even if they do not directly relate to troubleshooting.

## Safety

A lot has been written about personal safety. It is not my intent to cover this further in this book, except to say, do not get hurt or let others get hurt. It is not worth it.

Safety expectations have improved considerably since I started in this industry. Incidents that involve multiple deaths in our industry are typically the result of process safety failures. The material covered should directly improve Operation's ability to address these situations--specifically, how to make decisions that can potentially save lives or stop the incident early on.

There typically is not one single failure that causes these events. There is usually a series of failures, and if one of them had not happened the severity of the event could have been minimal.

Human error can always be determined as the cause. In almost every incident you could go back and find some point that if someone had taken a different action, the event would not have occurred. Pointing fingers is the least productive way of preventing future events (remember there are always three fingers pointing back at you). This approach can also lead to the investigation stopping and not learning about other key initiators of the event.

Process safety is successful when all the groups are working together--operations, engineering, instrumentation, control specialists, electrical, millwrights, pipe fitters and of course leadership (I likely missed a few). If there is animosity rather than cooperation amongst any of these disciplines, reliability is the first to suffer, and then process safety will be the next victim.

One of our largest safety nemeses is normalizing deviations. We become used to working around problems--flaky transmitters, unclear procedures, broken safety interlocks that have been bypassed or must be bypassed just to start the equipment up. Many of the events covered in the

Chemical Safety Board reviews have had similar problems occur in the past. We struggle with learning from the past.

Two books I found that gave some helpful insights into safety are, "Better Questions, Pre-Accident Investigations" by Todd Conklin[34], and as previously mentioned in the training chapter, "The Field Guide to Understanding Human Error "by Sydney Decker.

I will mention again how well the Chemical Safety Board does investigate events and they publish excellent videos. Try to learn from other's mistakes—it is much less painful. Create scenario-based training about these events. This develops pattern recognition and increases the chances of recognizing a similar incident in your process.

There are two specific types of process safety hazards that I will cover because of their severity and that they are common to most industries. Understanding these principles can help you prepare or prevent these events from happening. They may be covered procedurally in your unit.

### Case Study: Static Build-up

Although this incident involves a supertanker it applies anytime when equipment is cleaned. The SS Melo[35] was one of two supertankers built for Shell in the 1970s that had an explosion. The crude oil loaded on the tankers had sand entrained in the oil that separated during the voyage. To clean the storage tanks they used high pressure, almost boiling water in movable high-pressure jets (cleaning guns) that rotated vertically and horizontally, spraying the hot water at high velocity around the tank walls. This produced a static charge that ignited when the mixture of air and hydrocarbons dropped into the explosive range.

To prevent this problem the pressure jets were modified to prevent the static build-up, and in this case, they used the exhaust from their diesel engines to purge the combustible gases from the tanks.

## Cold Brittle Fractures

This is defined as the sudden and often catastrophic fracture of a material with little or no deformation of the material. This type of failure occurs rapidly with little or no warning such as bulging or leaking before the break. Brittle fracture occurs in a 'break before leak' mode rather than a 'leak before break' mode. This type of failure may release the contents of a vessel very quickly due to the size of the failure. Depending on the process, this may result in a large explosive vapor cloud that can be ignited by sparks from the failure or another ignition source nearby.

Carbon and other ferritic steels become more susceptible to brittle fracture as temperature decreases. Austenitic stainless steels (300 series), nickel-base alloys, aluminium alloys and copper alloys are essentially resistant to brittle fracture. The thicker the vessel wall is, the more susceptible it is to a brittle fracture. Thin-walled vessels will bend or crack relieving some of the stress.

Brittle fractures have occurred during temperatures above 0 deg. C (32 deg F). This is a concern when hydrotesting vessels in cold weather or using cold water. Determine the limits of the equipment before proceeding.

Here are the definitions for critical parameters:

**Maximum Allowable Working Pressure (MAWP):** The maximum pressure the vessel may experience. The typical guide is your PSV settings.

**Minimum Design Metal Temperature (MDMT):** The lowest temperature allowed at MAWP (to ensure the vessel remains ductile)

**Minimum Allowable Metal Temperature (MAT):** The lowest allowable temperature at a given pressure. At a pressure below MAWP, the metal is allowed to be cooler. Note MDMT = MAT @ MAWP.

**Critical Exposure Temperature (CET):** The lowest possible metal temperature due to process conditions at 40% of the MAWP.

The diagram below is the cold brittle fracture triangle that shows the three requirements for this type of fracture. The susceptible material was mentioned above. The sufficient applied stress is the process pressure on the vessel walls. The material flaw may be a metallurgical or mechanical notch such as a crack as is often found in weld or manufacturing defects.

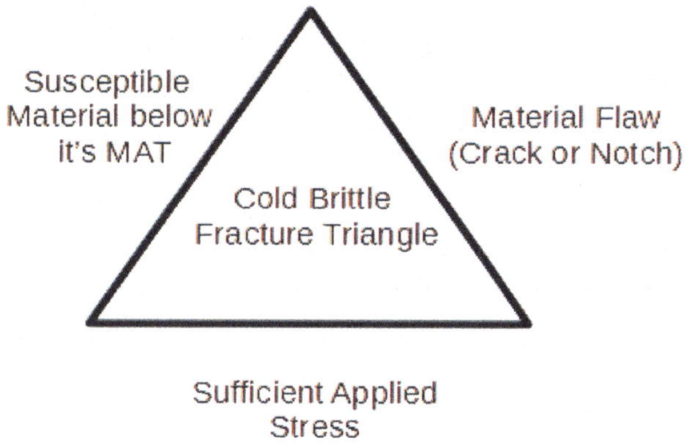


Since most vessels at normal operating conditions have the sufficient stress required as shown on the triangle, and no vessel is fabricated perfectly, it is assumed that during normal operation two sides of the triangle are always present. This is also a good reason to inspect vessels periodically and remove flaws.

The best way to avoid brittle fracture is to avoid completing the triangle. Do not allow the vessel to drop below the MDMT (Minimum design metal temperature) or MAT (Minimum allowable temperature) when operating.

If pressuring up a system, ensure equipment is warmer than MDMT. If the equipment is above 40% of MAWP (Maximum allowable working pressure) for carbon steel, the vessel may be at risk of brittle fracture if the temperature of the metal (not the process) is too low.

**Auto-Refrigeration**

Auto-refrigeration is defined as "the temperature change of a liquid, gas or vapor with a change in pressure that results in refrigeration". Any material that is normally a vapor at atmospheric pressure is susceptible to auto-refrigeration. The energy required to vaporize a liquid is referred to as the latent heat of vaporization of a substance under a given set of conditions. When equipment is depressurized some of this energy is drawn from the metal of a vessel or pipe.

Nitrogen in the presence of a hydrocarbon such as ethylene has the effect of lowering the temperature of the ethylene at a given process pressure. The lower the partial pressure the lower the boiling point of a hydrocarbon. Therefore, in a vessel with both liquid and vapor phases, the drop in partial pressure will result in a drop in temperature. When hydrocarbon freeing vessels, do not introduce N2 until liquids have been removed. Drain first, then purge. After O2 freeing equipment with N2, lower the N2 pressure as much as possible before introducing hydrocarbons. Purge the N2 from the system with the same hydrocarbon vapours as that of the liquids being introduced when possible.

**Recovery**

Always follow your company's procedures and policies when responding to abnormal situations. If they contradict what you find below start a discussion and feel free to let me know.

If the equipment is below the MDMT or MAT, the only side of the triangle that can be removed is the stress. De-pressuring would lower the stress if no liquids were present. If there are liquids present hold the pressure until all the liquids are removed and then depressure. If the chilling was caused by a loss of pressure, do not re-pressure the vessel until it has had time to warm up. Re-pressurization at this time would be completing the triangle. For carbon steels, once the pressure is below 40% of MAWP, the stress is considered below that which would complete the triangle. In other words, if a carbon steel vessel has been sub-cooled by auto-refrigeration and the new pressure is below 40% of MAWP, brittle fracture is no longer a concern. However, do not re-pressure until the metal temperature rises above MDMT. It would be prudent to consult with a Mechanical Engineer in this situation.

## Case Study: Auto-Refrigeration Near Miss[36]

In this incident, ethylene was backfilled from a pipeline to inventory two distillation towers. The second tower was required to be above 200 psi (1380 KPa) to prevent sub cooling; however, the pressure transmitters lower range was 200 psi. This transmitter was used for a controller and was on the main graphic. There was another pressure transmitter with a wider range on a secondary graphic. This transmitters history showed the tower was 22 psi while operations believed the tower to be at 200 psi as required.

The tower had two reboilers that used Quench water for heat and one back-up reboiler that used steam. Both Quench water reboilers were damaged and all three had frozen. The tower had been inventoried slow enough that it did not exceed the MAT for the pressure it was at. If one of the reboilers had been functioning the tower pressure may have increased enough to exceed the MAT for that pressure.

# Fuel-Rich in Fired Equipment

This is a dangerous situation where there is unburned fuel in the firebox of a boiler, furnace, or any other type of heater using combustion. Usually, the most risk is during the transition back to excess oxygen when there is a possibility of an explosion. This may also be called gassed up.

It can be a difficult scenario to deal with as it is somewhat counterintuitive. The first response seems to be that if you are short of air, you then add air. This applies until you are into a fuel-rich scenario. This would be the wrong response if you have next to 0% oxygen and the carbon monoxide is increasing. Reducing fuel flow at a rate of 2% per minute until excess oxygen is established has been recommended. The flue gas change-out rate of the fired equipment should be considered when calculating the appropriate rate of cutting fuel for that equipment. It is best to hold the airflow stable at this point to prevent any sudden changes.

Andrew Hawryluk submitted a paper to AICHE, Hazardous Flue Gas Mixtures in Furnaces Due to Fuel-Rich Combustion[37] that has some helpful information.

He mentions that in a fuel-rich condition at operating temperatures above 700°C (1292°F), the excess CH4 will partially oxidize and decompose into CO and 2 H2. This will result in more H2 present than CO, even when pure natural gas is used as fuel. The concentrations of CH4 and heavier hydrocarbons will be negligible at these temperatures. CO analyzers will likely not pick up the H2 and hydrocarbon analyzers would not indicate anything at these temperatures. There is a potential of up to 50% more energy available from the unburned fuel that the CO analyzer may not account for. It is best to determine what the analyzers in your facility will read and what their limitations are.

He also notes that the most destructive furnace incidents occur at light-off. Combustion from ambient temperatures causes a much larger relative temperature increase and a correspondingly larger overpressure. I would recommend a best practice or procedure be developed for your equipment on how to approach this condition. Clearing the area of all personnel and waiting to have your Field Operator respond until you have reduced the fuel to the point you have excess air again is recommended.

## Case Study: Positive Firebox Pressure Event

Late one night shift, two of us were training in our next operating areas, when a tired Panel Operator made a radio call to another Operator "Can you please check out the firebox draft on furnace nine?" That was all. There was nothing unusual about that call. It was a troublesome process heater from the early 80's. It was a balanced draft design that frequently needed attention by the outside Operator to keep a negative (and stable) firebox pressure.

Sometimes your spider senses tingle when there's no good reason for it. Was it the time of night? Was it the tone of voice? Whatever it was escapes me now, but both of us thought "Let's go see what's going on" even though the call was sleepily monotone, and the request was something we were accustomed to hearing on any given day.

I stopped in the control room on the way to have a look. The firebox pressure was off-scale high. "Yeah, the transmitter just died" was the answer when I asked why it wasn't just a bit positive, or unstable, like usual.

"Your induced draft fan damper is closed?"

"Yeah. It's been closed for days."

I pulled up a quick 24-hour trend (one of the quick-select options) and saw that he was right. It didn't make sense. How could a nearly fully loaded heater be running for days with the flue gas outlet damper shut? I zoomed out to a 7-day trend and saw that just like he said it had been there for days.

"There's fire coming out of the viewports!" came an excited call over the radio.

The outside Operator later told the story that he had meandered over to the bottom of the stairs in the cold night air. Tired and trudging along, he was staring at the ground as he approached the furnace to repeat the age-old pattern of TLC that the firebox draft transmitter needed to start working again. As he grabbed the handrail to go up the stairs to climb the side of the furnace, some unknown sense made him stop after only going up the first step. "Why stop?" "Look up" He looked up, and a few feet further up the flight of stairs a blue methane and hydrogen flame was coming out of the view port door extending about 3 feet across the walkway and stairs. The furnace was again proving its nickname "fire-breathing dragon" was appropriate

"Well, stay away from that furnace--the firebox pressure is off-scale high," the Panel Operator replied. The right information doesn't always arrive at the right time.

We sounded the "Alert" to let other units on the site know that something was up.

Three of us scoured the panel graphics and instrumentation, pulling trends. It was quickly and clearly obvious that whatever the cause of the high firebox pressure, the firebox was VERY fuel rich, and needed to be very carefully transitioned back. The excess O2 was nearly 0%, the carbon monoxide (indication of incomplete combustion) had spiked high, and heater temperatures were falling across the board. The automatic control systems were adding fuel to this already fuel-rich firebox to try to raise the temperatures back up to setpoint and adding air via the forced draft fan damper.

If left alone, this would lead nowhere good. Tripping the fuel supply to the furnace would be just as dangerous. There was a large volume of fuel above its auto-ignition temperature, more entering continuously, consuming the air from the forced-draft fan that was still entering the firebox. To suddenly stop the fuel flow would allow the air to be un-consumed leading to a sudden availability of excess air. Super-heated fuel plus lots of air equals something much worse than we already had underway. We were trapped in a tango we couldn't just stop.

The age-old tradition of fuel-rich firebox response was put into motion. The forced-draft fan and induced draft fan dampers were set to manual so that airflow would not increase. Most of the fuel gas control systems were turned off, and the lowest level controller was set to ramp fuel gas flow down at 10% each ten-minute period.

"This fan damper is closed!" came another call over the radio.

"What are you doing on the top of the furnace? That furnace is not safe to be around right now." the control room called back. "Please come in to the control room." Sometimes you need to position things or people so that they are manageable without extra effort. "We're bringing the furnace down, and we're going to need to talk about how we're going to do it without going on the furnace."

Normally, shutting down this old furnace involved a lot of manual steps in the field to shut it down. We had to transition the firebox back to excess O2 by reducing the fuel flow down to minimum slowly because a rapid transition could be catastrophic. The firebox was about 3 stories tall and 30 feet wide. That's a lot of fuel above its auto-ignition temperature just waiting for fresh air to ignite. If there was another sudden change when trying to transition back to excess O2, anyone in the area would be at risk of injury or worse. We landed on a plan in minutes and then had to wait uncomfortably as we reduced the fuel gas flow at a rate of 2% per minute.

"This furnace is getting really hot!" came another call on the radio.

"Please come into the control room," replied the Panel Operator.

"I'm going to get him and see for myself," I decided.

When I got to the outside Operator, he was standing on the ground near the furnace. There were indeed blue flames coming out of each of the roughly 3-dozen firebox inspection ports. Where coil hangers and supports penetrated the furnace wall, there was an opening sealed by a fire-proof bag. Superheated fuel was exiting most of these locations as well and instantly igniting in the fresh air, making flames around each bag. The structural steel that held up the furnace was starting to glow a dull red in some areas. This furnace wouldn't last much longer.

We both went back to the control room. The fuel was, by now, nearly down to minimum, and within a short time transitioned just barely back to excess O2, and we tripped the fuel to the furnace.

We continued air flow to cool the very hot furnace. After about an hour, all instrumentation said the furnace was safely offline and now cooled off.

We went out to survey the situation. Looking inside the furnace was like looking at a volcanic moon in a science fiction movie. Every refractory firebrick was cool and black on the outside surface, but inside each crack was a red-hot glow like a river of lava. The heat had soaked through the firebrick to the layers of insulation underneath. The oxygen-starved flames coloured the normally white firebricks many shades of green, purple, orange, and black. Cables and tubing bundles were melted. The structural steel was not permanently damaged. Repairs were needed to most of the soft goods attached to the furnace or near openings, but nothing catastrophic occurred.

The incident investigation discovered that the induced draft fan damper had become stuck at about 50% open, a small offset in the feedback caused the control system to try to close the damper the tiniest bit. But because the damper was stuck, the control system continued to try to close the damper for days until the commanded position was 0%.

A recent replacement of the damper positioner had a new setting that was not very well understood. The factory default was if the current signal was below a certain threshold for a certain duration, the positioner would turn itself off. When the positioner went to sleep, it fully vented the air from the actuator and the additional force was sufficient to break the damper free, causing it to suddenly go from about 50% open to a fully closed position.

**In hindsight:**

- How did the command to close the induced draft fan damper slowly over days not get noticed by multiple Panel Operators over many shifts?

- How did the command to have the damper at 0% position for over a week not get noticed by multiple Panel Operators over many shifts?
- Why was there no alarm on the damper being below some minimum position while the heater was fully online?
- Why had the minimum travel stop (a physical bolt) on the damper been adjusted to basically fully closed?
- Why were people, including myself at one point, inclined to go out into the line of fire near a piece of equipment that is in a dangerous condition?
- Sometimes the obvious answer (that the damper was shut) is there the whole time but can be overlooked if you only compare to historical trends instead of basic principles.

***Contributed by Awny Betts***

## Simulators

There has been a lot of discussion about whether the cost of simulators is worth what they offer. Unfortunately, we are not able to look into a crystal ball and predict that a simulator will prevent a catastrophe or a plant outage. Dan Heath discusses this challenge in his book "Upstream: The Quest to Solve Problems Before They Happen".[38] He discusses the three barriers to upstream thinking: problem blindness, a lack of ownership and tunneling.

I do not think a lack of ownership is a problem in this area. Problem blindness could be an issue. One of the most difficult things for someone that has not sat in the hot seat is to understand the amount of pressure and demand that can be made on a Panel Operator. The analogy that I like to use is to practice being a NASCAR race car driver going 60 mph. You can become proficient at shifting gears, some acceleration, how to steer and braking. How do you think this driver will fare on race day when speeds exceed 200 mph, taking tight corners with several other cars inches away? A NASCAR driver would have to rely heavily on pattern development to be able to accomplish this.

A NASCAR driver would be unable to have the time for a rational decision approach that may look something like this: If I use my tires coefficient of friction, the weight of the car, my speed, and the angle of the curve I should be able to take the corner at 140 mph and stay at least 6 inches away from the car beside me. A Process Operator may not need to develop the motor skills to drive a race car but they will have to develop the cognitive skills to manage a process where the problem solving and decision-making has accelerated to 200 mph.

When the simulator program was first introduced at our site the original trainer had noticed the difference in skill level between some of the

Operators and that there were some surprises. Confidence is not necessarily a good indication of competence. We can be lulled into a false sense of security if the run times of the process are 99%. This could be an indication that the Operators are performing very well. This could also be an indication that the Operator's skills are not being challenged and are deteriorating.

When neural connections are not used, they are assigned to something else. Advanced computer controls can also have an impact on an Operator's skill level. They do have an advantage as they can be continually tweaking the process to optimize production.

A tunnelling barrier could have an impact on getting to the point of deciding on starting a simulator program. The individuals that are required to make this decision or to promote the idea are typically taxed by keeping the process running and dealing with the day-to-day challenges. It typically takes someone in leadership to realize the value and to promote it.

Money is usually an issue, and it will be scored against other projects. There are always cycles in industry, and when the prices of the product are increasing, that is the best time to promote a simulator project. The other time would be after a major event (sad but true).

The Texas City Refinery now has a simulator, and this was one of the CSB recommendations. I have also been told that BP was developing simulator programs in their other facilities. I am not trying to suggest that it is Operator error that causes these large events. There are typically several things that contribute to these events. Well-trained Operators certainly are one of the solutions.

We can look at some factors that can be used in the decision of starting a simulator program:

- I have mentioned reliability and how much interaction the Panel Operator has with the process. If the facility is only shut down every five years then what will the competency of the Operators be for this task?
- The complexity of the process.
- The cost of losing production if the plant trips. The length of time to start back up should also be factored in.
- Environmental impacts of flaring
- Community impacts of flaring or of a release. How close are the neighbours and how toxic is the product?
- The turnover of operations. How fast is experience walking out the door?
- Would you like to test software upgrades and control system changes before implementing them in the unit?

The fidelity or how close the simulator model is to the actual unit is another consideration. There are basic simulators that will allow Operators to develop some skills interacting with a console and a general understanding of the process. There are now models that are considered high fidelity but are generic to the same process in several facilities.

The simulators I have worked with are modelled specifically for that unit. The consoles are similar and the same graphics that are used in the plant are on the simulator. The control systems and safety logic are the same. The closer the simulator is to the process the better the pattern recognition skills of the Operators will be, and the more skilled they will be at problem-solving.

One of the keys to maintaining a simulator is developing a respectful relationship with your modelling engineer. Patience is required at times because of the complexity of our processes and the number of data points involved. Documenting any of these issues during developing is critical. I was fortunate to work with Bob and Kiran from CORYS Simulation.

Here is a list of the benefits of a simulator:

- Instructors can develop and validate the competency of the Panel Operator's ability to respond to emergency procedures and process upsets.
- Engineers can test software updates for the control system on the simulator before downloading it on the live system.
- Plant personnel can test changes to the controllers and advanced computer control systems before implementing them in the live system.
- Instructors can train the control engineers and control system specialists on the simulator first.
- The simulator allowed my plant to develop and test automated procedures for known upsets. It is very difficult to develop these without a simulator to run the numerous tests required to make these work, unless you have an unreliable process that gives you many opportunities to test them. Automated procedures free up the Panel Operator to respond to additional problems that may occur (leaks) and contact other impacted units.
- Plant personnel can develop strategies to reduce flaring during start-up and upset conditions. This has obvious environmental benefits and can help with being a good neighbour by reducing flaring noise and duration. My plant was able to significantly reduce flaring amounts by testing new strategies on the simulators.

This is a link to a paper I presented at the 2012 AICHE, Ethylene Producers conference with Roger Young and Michael Guaraldo as coauthors.

Operator Training Simulators, Experience for Training, Certification and other applications[39]

# Start-up

Starting up a facility can be one of the most challenging and exciting tasks an Operator is required to do. This can be from a new facility or an existing one that has just gone through a turnaround and possibly a trip from a malfunction.

I have included this because there can be so many problems that require troubleshooting and decision-making. I have been involved in approximately 15 start-ups and almost all of them had unexpected secondary issues. These could be from transmitters or controllers not lined up, transmitters failing, pumps not started, or seals leaking and process leaks to the point the plant needs to shut down again.

The key to a successful start-up is preparation. I will not go into commissioning here as that could be a book in itself.

The following is a list of key areas:

**Leak checks -** Every flange that has been opened should be checked. Pressuring up the equipment with nitrogen and using helium is a very effective method.

**Line-up checks -** Use 11 x 17 copies of the P&IDs and organize them into the systems with the boundaries marked on them. Have some thin boards and a sheet of plastic cut to size and use large paper clips to attach them. These will be used in the field and marked with highlighters as each piece of equipment is checked. Check that every transmitter in is service and a visual check of the sensing lines. Stroke check all the control valves and look for smooth operation. Make a note of any valves that are still closed. PSVs have been lifted and start-ups delayed because manual valves have been closed and a compressor or pump is started up. Look that all the drains are closed and any possible way that the process can escape. Drain caps or plugs should be installed on hydrocarbon or hazardous services.

Look for nuts missing on studs, pipe supports not on their shoes, spring hangers within range. Ensure heat tracing is in service--this includes glycol, steam, or electric. This is important even if the ambient conditions are warm as some facilities have cryogenic systems that freeze sensing lines and hot services can cook non-circulating glycol and cause flow restrictions or block the flow.

**Safety logic -** All instrumentation involved in safety systems should be checked including the actions required of the logic. Typically, this requires the instrumentation technicians to force a signal from a transmitter to test the action of the logic.

**Exchangers -** Exchangers will have a temperature differential limit, and if this is exceeded the head may leak into the atmosphere. There may also be limits on the rate of changing the temperature. When starting flow through a fixed tube exchanger the best practice is to start flow through the shell side first. In this way, the tubes will change temperature with the shell at the same time. This applies to cold service as well as hot. Also, when shutting the process down reverse the order and remove the process flow from the shell side last when possible.

In the diagram on the next page, the flow is started through the tubes only. If the process is hot the tubes will expand and the shell will stay the same, forcing the tubes to either bend the tube sheet or break the seal of the tube to the tube sheet. These conditions can cause the gasket to leak to the atmosphere or process gases leaking from the high-pressure side to the low-pressure side contaminating or possibly over-pressuring the other side.

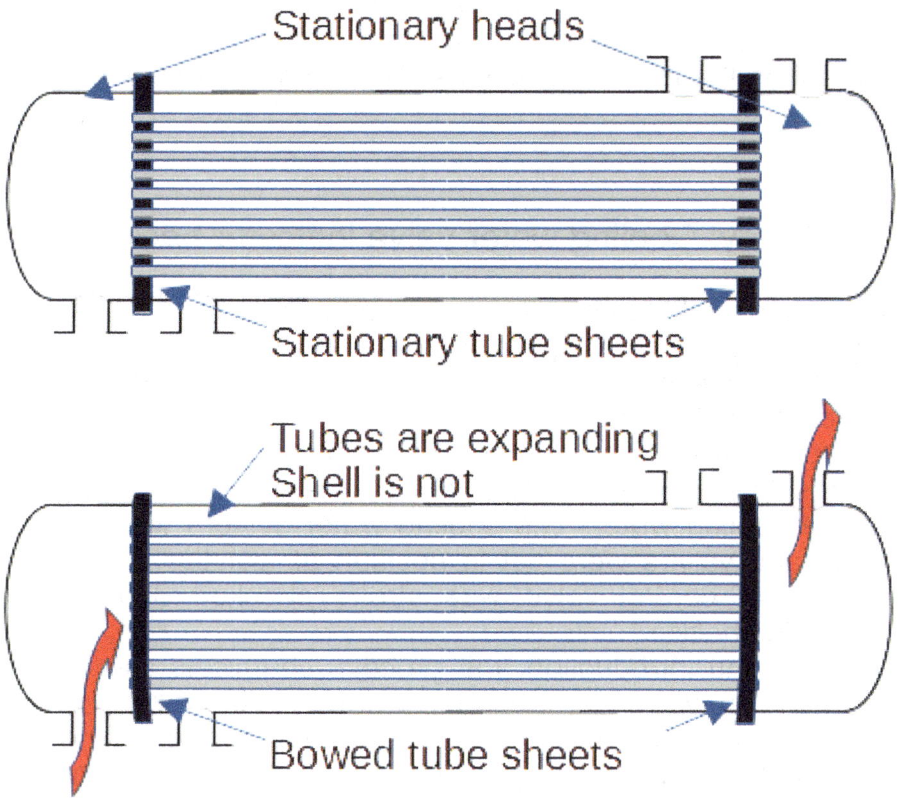


**Start-up organization -** Responsibilities should be assigned well before to allow operations to prepare and review procedures. If a simulator is available, they can practice start-up a few days before. This is an excellent way to validate procedures and ensure understanding of the interactions of the different systems.

Maintenance support can be identified ensuring at least one person for each skill is available, Instrumentation, Analyzer Technician, Electricians, Millwrights, Pipefitters (leak response), Process Engineer, Control System Engineers/Technicians.

**Start-up day -** Make a list for the night shift of the condition the process needs to be in on start-up morning. Do any compressors need to be running or on slow roll, towers on total reflux, inventory in any vessels or towers or valves in a certain position?

If your process requires more than one Panel Operator during start-up it is helpful to divide up the panel beforehand, to avoid two Panel Operators making moves on the same controllers. The split can be just halfway down the process. Another approach that has been used on a complex process that requires three Panel Operators on start-up day is called the "feed-forward" technique. The panel is split into systems with each Panel Operator taking turns as these systems are brought online and feed to the process is started. The Panel Operator that brought the first system online then brings the fourth system up and so on. When a system is first brought online it requires the most interactions. This technique helps to split up the workload.

## Case Study: Cooling water valve indicator backwards

This scenario could happen in many processes.

During start-up after a major turnaround in the plant, we were initiating feed into the process vessels. The chemical introduced in the system generates exothermic heat.

The heat exchanger has clean water on the shell side and cooling water on the tube side. The clean water picks up the process heat and sinks the heat into the cooling water. The cooling water system was commissioned beforehand, and valves were set in position as per the start-up checklist. The temperature control valve was stroke checked and the operation verified via field check.

During the feed introduction, the process chemicals started to release heat and the clean water loop began to warm up. After about 10 minutes of the feed introduction, the temperature of the loop kept creeping up with no delta T across the heat exchanger. We had a high temp alarm on the loop and at that point, we pulled the feeds and did further troubleshooting. The clean water pump pressure and flow, TIC field position VS DCS output, and temperature transmitters were identified as functional. The heat exchanger outlet temperature was above the boiling temperature when compared to cooling water system pressure. It was finally noticed that the cooling water outlet valve "indicator" was installed backward which led to the valve being closed internally (butterfly valve).

**Lesson Learned:** Manual valves should be checked diligently and verified against delta P across the exchanger. Temperatures should be monitored closely during start-up to prevent runaway reactions scenario."

**Contributed by Taj Bajwa.**

This incident also reminds me of one of the first lessons I was taught as an Operator: "righty-tighty, lefty-loosie". I hope this applies to every valve in the industry and it means to turn the valve clockwise, or to the right, to close it. Possibly this problem may have been picked up with that in mind, although I do not have all the details.

Here are two incidents investigated by the Chemical Safety Board. I will not repeat the learnings from them. Below are the links to the material instead. They are both well worth looking into. You can also search for them on the Chemical Safety Board web page at www.csb.gov/videos/.

**Anatomy of Disaster (BP Texas City Refinery)**

This is one of the more discussed incidents in our industry and has several key learnings from it. Here is a link to the video "Anatomy of Disaster" [40]by the US Chemical Safety Board.

There is a book written by Andrew Hopkins called "Failure to Learn, the BP Texas City Refinery disaster". He does an excellent job of breaking down the whole event all the way from executive leadership to the decisions of the Operators on shift that day and the previous shift.

**Brazed Aluminium Heat Exchanger (BAHX) fails during shutdown**

This event was investigated by the Chemical Safety Board and the report can be found on their web page. Enterprise Products Midstream gas plant at Pascagoula, Mississippi.[41] This event occurred during a plant shutdown; however, similar conditions can be present during a start-up.

# If I were allowed to design graphics

Now that we have covered how we make critical decisions with time constraints, would it be worthwhile to design the graphics that helped us accomplish this as quickly and efficiently as possible? There has been a trend to develop multi-level graphics with the theory that when we have a problem, we will identify it from a basic overview (sometimes very abstract) and then drill down two or more levels to finally arrive at the correct amount of detail to solve the problem. This whole thought process is seen as driven by an alarm. Most of the time this may be the case; however, there are times the first alarm that comes in is from a secondary system impacted from the system where the actual event originated.

The other challenge with this idea is that this is not how we make decisions, especially when there are time constraints. As we covered earlier, 90% of the time one course of action is determined. This is accomplished with pattern recognition. The question is then: How do we help Operators pick up quickly on what a problem is?

I do not intend to be critical of the engineers that developed this system. That is how they see the world and it reflects the way they problem solve. They typically have sufficient time to work their way through a problem one step at a time. The awareness of pattern recognition and the Recognition Primed Decision model is just making it into our industry, even though Dr. Gary Klein wrote "Sources of Power" 22 years ago.

The first two examples are for level controllers. Deviation from the setpoint is indicated by increasing the amount of colour either below or above the setpoint line. The different colours represent the severity of the deviation with blue as a low priority, orange as high priority, and red as emergency. These could be configured depending on the process and criticality of the level indication. An audible alarm or alarms can be set for any point as well and not dependent on the colour. They also readily indicate if the PV is deviating high or low from the setpoint. Therefore, an Operator can easily determine from a glance if there are PVs deviating and in what direction. The two objects beside the level indicators are for temperature (hopefully obvious). The two below them are for pressure and the rectangle beside them is for flow.

The coloured indication does not have to be the full range of the transmitter. This may take away some of the resolution near-normal control if the transmitter range is quite large. A small tick could be added to indicate if it is a full range, and another added if it is at the pressure safety valve setting as indicated on the pressure object.

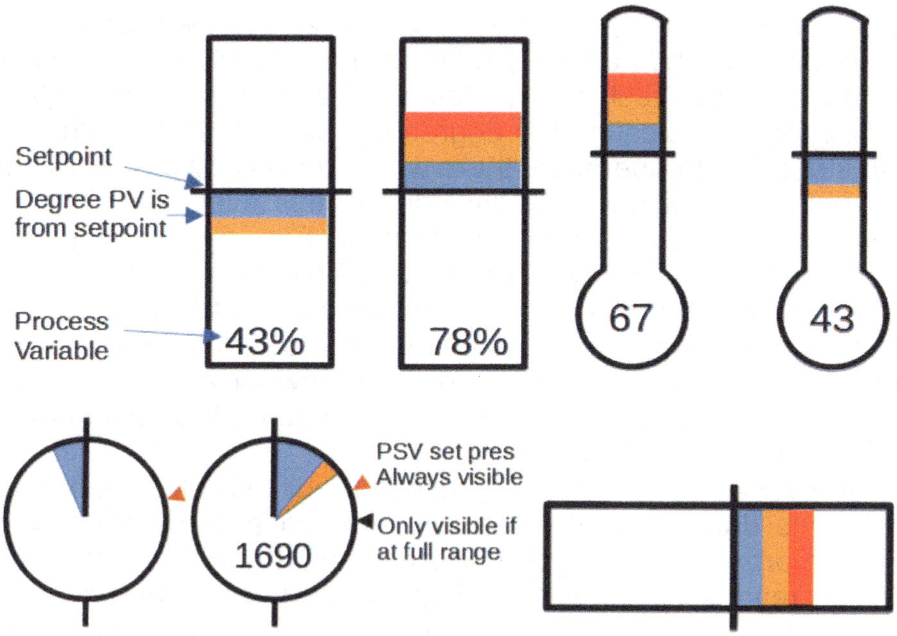


The next two examples are for a flow control valve with the output indicated with a pie chart at 25% open. The outer circle could be removed to only show the valve output, or it could be used as a deviation for the output. I believe the pie graph for an output is easier to recognize the actual output, and it is more intuitive as it is representing how much area the process fluid or liquid would have to flow through. Contrast this to the traditional valve object that depicts it from the outside as shown to the far right.

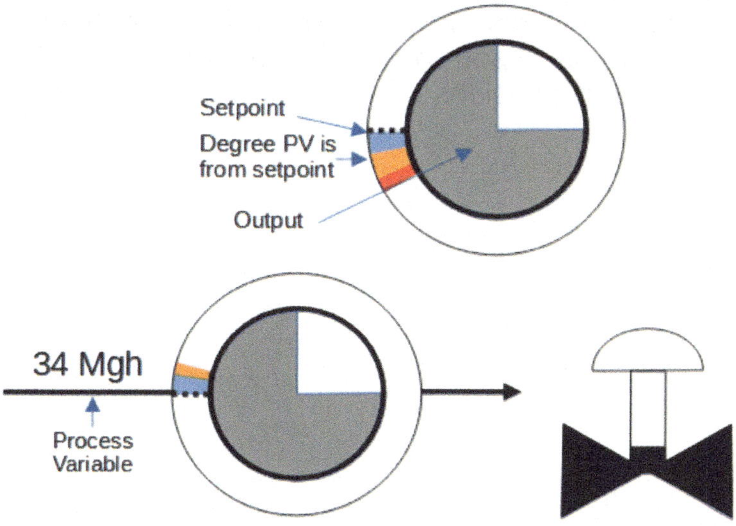

The diagram on the next page is a partial overview of a process to provide an example. As I am only proposing two levels of graphics the next level would have the rest of the indicators and PVs. With a glance, the colour is picked up indicating the degree of deviation and whether it is high or low. In steady-state operation, there would be very little colour showing.

The whole process would be included in this overview. This may take one large monitor or more than one monitor. For key parameters, a trend could be embedded in the graphic as well. The layout of the next level of graphic should match the overview as closely as possible to aid in pattern recognition when moving from the overview to the detail level. The size of the indicators can be adjusted depending on how far away the monitors are from the Panel Operator. They are the main feature; the equipment shapes are just there for landmarking.

With an overview of the whole process readily visible, it will be glanced at many times during a shift. This will develop a comprehensive mental model of the process and a good sense of what normal is. Not only will a deviation be readily visible, but the impacts on the rest of the process will also be obvious. This will lessen the likelihood of an alarm being lost in an alarm flood as it will be graphically visible with the colours progressively indicating the severity of the problem.

It is easier to lose track of multiple problems when drilling down into a detailed graphic. This can happen when you are dealing with a problem and another alarm comes in that takes you to another graphic. While in this graphic it can be easy to forget about the previous issue. An occasional glance at an overview will re-establish the problem areas. Alarm summaries do not lend themselves to pattern recognition. They are a constantly changing string of events that take time to read through and establish what they mean. A graphical overview makes it easier to perceive the connection between events. As discussed in the pattern recognition section, there is less cognitive load and decisions can be made faster with the use of pattern recognition.

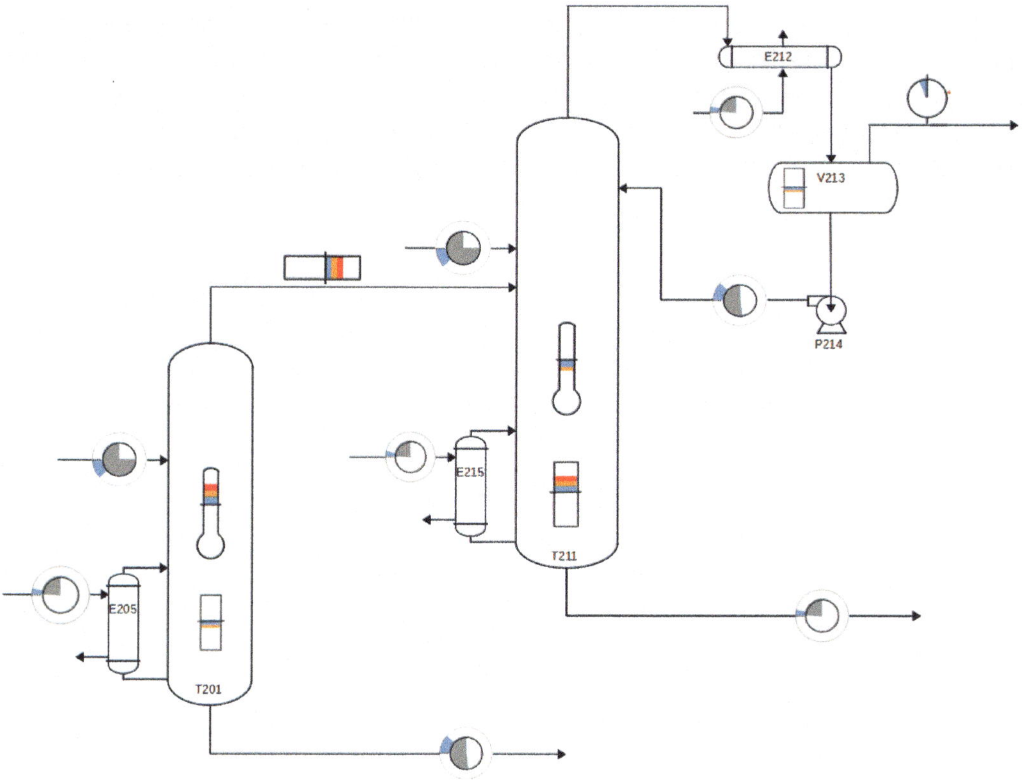
E212
V213
P214
E215
T211
E205
T201

This is the detail graphic with setpoints, and process variables added. Additional indicators not involved in control only have the PV indicated. There can be other objects developed with this theme in mind and additional features added to these objects. A trend window could be added to the bottom of the graphic with a detail window that would display all the information of each point when selected.

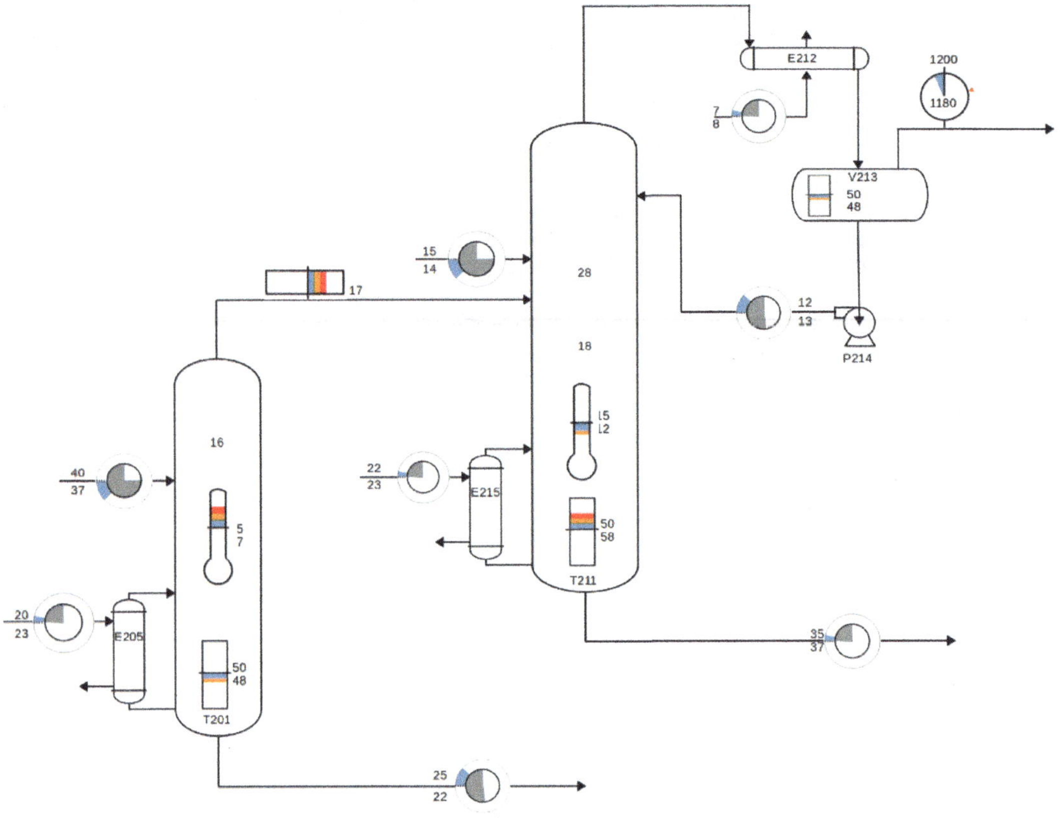

# Acknowledgments

This book came about from five years of research with the COP (Center for Operator Performance). The largest influence came from the five research projects I championed with Dr. Gary A Klein and Joseph Borders. John Schmidt was also involved in developing some of the training tools developed for the COP. I also championed two projects with Dr. Beth Blickensderfer. The second one of those was also with Dr. Joseph Keebler.

It is difficult for me to give credit for every piece that these individuals shared with me. That information has become entwined with all the other information and has had a huge influence on my understanding of how operators can make critical decisions.

This research was also used to develop a Troubleshooting course in the Petrochemical site I worked in. This was completed with my fellow Training Specialists Ben Courtier, Cliff Roy, and Roger Young. Ed Noble and Fred Ruether were also involved in some of these discussions on Troubleshooting.

These experiences have been a catalyst for additional concepts and branched off into research from additional authors.

The largest influence and what generated one of those Aha! moments for me was Dr. Klein's book "Sources of Power, How People Make Decisions". The concepts in this book address how people make critical decisions when time is limited and describes the recognition primed decision model. Dr. Klein has researched Fire Fighters, Air force Pilots, Police, Nurses, and more recently petrochemical operators. His research is directly applicable to help process operators. I am very grateful Dr. Klein allowed me to use anything he has published. I hope you found this information as useful as I have and share it.

Joseph Borders' insights, project skills, and patience were also key ingredients to the success of the research we completed. I am quite looking forward to Joseph's career as he is multi-talented and will bring innovations to the field of cognitive psychology.

I had several discussions with Dr. Beth Blickensderfer that lead to new insights and paths to explore. Her perspective of our industry was quite beneficial to the COP projects she worked on.

I have also had the pleasure to work with some very talented people who have demonstrated important values for working in our industry. Although there are many, I would like to acknowledge two that stand out, Mark Williamson and Randy Saunders. They are two masters in their field and have patiently explained concepts to me on many occasions and took the time to discuss troubleshooting approaches with me.

In my role as a Training Specialist, I conducted training with a high-fidelity dynamic simulator for the last 18 years. We also used the simulator to conduct three research projects for the COP. Observing this research and reviewing the findings of our researchers proved very helpful in understanding how operators worked through problems. I was very fortunate to have the cooperation of the operators at the facility I trained at. They willingly participated in these research projects with cognitive psychologists observing them work through malfunctions on our simulator. I am honored they trusted me with this arrangement.

Several conversations and insights were provided by Tessa Van Oene that had a large influence on the window of clarity section.

I believe the case studies are a valuable tool in sharing experiences across our industry and I was very fortunate to have several excellent contributions and would like to thank everyone that sent in a study.

I would also like to thank everyone who reviewed this work: Dr. Beth Blickensderfer, Barb and Dave Baker, Randy Saunders, Lisa Via, David Strobhar, Clary Michaels, Joseph Philip and Christine Goetz.

# End Notes

[1] https://www.istockphoto.com/photo/view-of-the-synapses-brain-connections-neurons-and-synapses-communication-and-gm1219443556-356709698

[2] (Adapted from Norman, D.A., "The Design of Everyday Things", Basic Books, 1998, P14-15. Also, thanks to Ben Courtier for expanding on the concept)

[3] https://www.psychologytoday.com/us/blog/seeing-what-others-dont/202101/the-mental-model-matrix

[4] Gary A Klein, Sources of Power: How People Make Decisions, (The MIT Press, 1998)

[5] Ray Kurzweil, How to Create a Mind, The Secret of Human Thought Revealed, Penguin Books; Illustrated edition (Aug. 27, 2013)

[6] David Eagleman, Livewired, The Inside Story of the Ever-Changing Brain, Anchor Canada (May 11, 2021)

[7] Lisa Feldman Barrett, How Emotions Are Made: The Secret Life of the Brain, Harper Paperbacks; Illustrated edition (March 13, 2018)

[8] Gary Klein, The Power of Intuition, Currency by Doubleday, 2003

[9] https://www.youtube.com/watch?v=C8k-lrJrldw

[10] Brain image by Gordon Johnson, Pixabay

[11] https://www.psychologytoday.com/intl/blog/seeing-what-others-dont/202005/the-second-wave-critical-thinking

[12] Gary Klein, Performing a Project Premortem, (Harvard Business review magazine, September 2007)

[13] https://www.youtube.com/watch?v=XDJcQBkcrus

[14] Joseph Ledoux, Anxious: Using the Brain to Understand and Treat Fear and Anxiety, Penguin Random House LLC, (2015)

[15] https://emcrit.org/wp-content/uploads/2017/05/PEP-under-Stress.pdf

[16] Hendrie Weisinger, J. P. Pawliw-Fry, Performing Under Pressure: The Science of Doing Your Best When It Matters Most, Crown Business, 2015

[17] Colin M. Shapiro, M.D., Ph.D., Ronald J. Heslegrave, Ph.D. Joanne Beyers, M.A., RD, Louise Picard, B.Sc. N., M.Sc.(A)., Working the Shift, A Self-Health Guide JoliJoco Publications, Inc. 1997

[18] www.operatorperformance.org

[19] https://mca-marines.org/blog/gazette/designing-good-tdgs/

[20] Gary Klein, Streetlights and Shadows, Searching for the Keys to Adaptive Decision Making, (MIT Free Press, 2011)

[21] Sydney Dekker, The Field Guide to Understanding Human Error, (CRC Press, 2014)

[22] https://www.psychologytoday.com/us/blog/seeing-what-others-dont/201809/cognitive-coaching

[23] https://operatorperformance.org/

24 David A. Strobhar, PE, Human Factors In Process Plant Operation, (Monentum Press, 2013)

25 Mark Goulston, How to Deal with the Irrational and Impossible People in Your Life, (AMACOM, 2015)

26 https://www.chemicalprocessing.com/articles/2021/how-good-are-techniques-for-handling-abnormal-situations/

27 Josh Waitxkin, The Art of Learning, An Inner Journey to Optimal Performance, (Free Press, 2008)

28 Carol Dweck, Mindset: The New Psychology of Success, (Ballantine Books, 2007)

29 https://www.psychologytoday.com/us/blog/seeing-what-others-dont/201605/mindsets

30 https://www.youtube.com/watch?v=H-XfCl-HpRM

31 Daniel H. Pink, Drive, The Surprising Truth About What Motivates Us, (Riverhead Books, 2011)

32 David Eagleman, Livewired, The Inside Story of the Ever-Changing Brain, (Vintage, 2021)

33 Ronald R. Morgan, Judith A. Ponticell, Edward E. Gordon, Enhancing Learning in Training and Adult Education, (Library of Congress Cataloging-in-Publication, 1998)

34 Todd Conklin, Better Questions, Pre-Accident Investigations , (CRC Press; 1st edition, July 6 2016)

35 http://philpott.de/Pdf_book/SS%20Melo%20%28Shell%20Tankers%20UK%29.pdf (page 64)

36 Autorefrigeration Near Miss Incident at Equistar Chemicals, a Lyondell Company, Michael J. Kennedy

37 Paper No. 38b, 2005 AIChE Spring National Meeting, April 2005, https://www.aiche.org/conferences/aiche-spring-meeting-and-global-congress-on-process-safety/2008/proceeding/paper/87f-hazardous-flue-gas-mixtures-furnaces-due-fuel-rich-combustion

38 Dan Heath, "Upstream: The Quest to Solve Problems Before They Happen" [Dan Heath, Simon and Schuster, (2020)

39 https://www.aiche.org/conferences/aiche-spring-meeting-and-global-congress-on-process-safety/2012/proceeding/paper/50e-operator-training-simulators-nova-chemicals-experience-training-certification-and-other-0

40 https://www.csb.gov/videos/anatomy-of-a-disaster/

41 www.csb.gov/file.aspx?DocumentId=6106

Made in the USA
Monee, IL
02 November 2026

67377577R00083